AN UNDISCOVERED SCIENTIFIC CODEX

MUKHTAR HASSAN

Copyright @2021 by Mukhtar Hassan

All rights reserved. No part of this book may be reproduced in any form or by any electronic or mechanical means, including information storage and retrieval systems, without permission in writing from the publisher, except by reviewers, who may quote brief passages in a review.

This publication contains the opinions and ideas of its author. It is intended to provide helpful and informative material on the subjects addressed in the publication. The author and publisher specifically disclaim all responsibility for any liability, loss or risk, personal or otherwise, which is incurred as a consequence, directly or indirectly, of the use and application of any of the contents of this book.

WORKBOOK PRESS LLC
187 E Warm Springs Rd,
Suite B285, Las Vegas, NV 89119, USA

Website: https://workbookpress.com/
Hotline: 1-888-818-4856
Email: admin@workbookpress.com

Ordering Information:
Quantity sales. Special discounts are available on quantity purchases by corporations, associations, and others.
For details, contact the publisher at the address above.

Library of Congress Control Number:
ISBN-13:   978-1-956876-63-5 (Paperback Version)
                    978-1-956876-64-2 (Digital Version)

REV. DATE: 02/12/2021

In our times it seems we cannot keep up with newly discovered scientific findings. In every field the discoveries seem to reinforce the notion that the past is no longer relevant, and there is nothing to gain from beliefs which now seem superstitious. However, the intention of this work is to analyse findings that had been discovered recently – only to realise that they were alluded to in a codex which was discovered centuries earlier…

Contents

Astronomy...	3
Atmosphere..	4
Red Rose Nebula..	4
Gravity...	5
Big Bang..	7
Stardust..	8
Speed of Light...	10
Biology...	20
Worker Bees are Female............................	21
Hearing before Seeing...............................	21
High Altitude...	22
Fingerprint...	25
Three Dark Stages.....................................	25
Made from Water.......................................	26
Prefrontal Cortex.......................................	26
Bones before Muscle.................................	26
Gender Determination...............................	27
Evolution...	27
Miscellaneous..	30
Flying	31
Barrier...	31
Iron...	32
Bounce and Crack.....................................	33
Relativity..	34
Quantum and Light Coherence...................	35
Ouzo Effect...	37
Number 19..	39
Mathematics...	40
The End..	44
Citations..	48
Bonus..	49

An Undiscovered Scientific Codex

Astronomy

Atmosphere

The discovery of the atmosphere is credited to two men: Blaise Pascal and Florin Perier – **in 1648**. Before then, humans had no idea that there is a layer of gases that surround the planet, which protects life from harmful effects such as ultraviolet radiation and meteors. So how could it be that the following statement was expressed centuries earlier?

And we made the sky a protected ceiling, but they, from its signs, are turning away

Red Rose Nebula

The Hubble Telescope was launched into space in **1990**. Whilst it is not considered the first to be launched into space, the specifications of it is what gives the consideration as being the first truly versatile telescope to be launched into orbit by humans. In other words, the images that it had subsequently captured are the first to <u>ever be seen by humans</u>. Some of the images captured were of Nebulas. Nebula is the name given to cloud-like masses of gas in space. Before they become Nebulae they are stars, and since these stars are very large, they release gasses into space because of their high internal pressure and temperatures. These eruptions of gas are very large and fast. These gasses then coalesce to form a gas cloud, with a temperature of more than 15,000 °C.
One type of nebula is known as the "Rosette Nebula" because of its resemblance to a rose. The Rosette Nebula is also a vast cloud of gas and seems to have an area five times greater than that of the full moon. Its' true diameter is estimated at 130 light years. A team led by Leisa Townsley, a senior Penn State University researcher in the field of astronomy and astrophysics, examined the Rosette Nebula using the Chandra X-ray telescope. They imaged hundreds of stars in the Rosette Nebula and determined that by crashing into one another stars produce gas at temperatures of 6 million degrees. Townsley describes what she saw: "a ghostly glow of diffuse X ray emission pervades the Rosette Nebula and perhaps many other star-forming

regions throughout the Galaxy. We now have a new view of the engine lighting the beautiful Rosette Nebula and new evidence for how the interstellar medium may be energized".

As the image was captured in the 90s, it would have been humanly impossible for anyone to have known how a Nebula would have looked, let alone use a metaphor – a rose – that beautifully describes its' appearance. Then how could the following statement have been made centuries earlier?

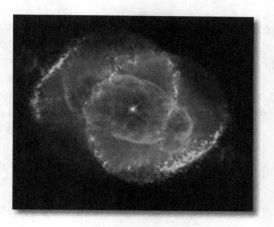

When the Heaven shall be cleft asunder, and become rose red, like stained leather.

Gravity

If there was no gravity than the earth would not be able to orbit the sun, all the different elements the earth is composed of such as the atmosphere, soil and oceans for example would all drift off into space. Isaac Newton is credited for discovering gravity by having an apple drop on his head, but Einstein posited that it is space which is pushing itself upon earth. He stipulated that the effects of gravity upon objects affects space-time around them. An analogy he presented is 'just as a boat sailing through the ocean produces waves in the water (ripples) moving masses such as stars and black holes, produce gravitational waves in the fabric of space-time. On February 11th **2016**

David Reitze, the executive director of the Laser Interferometer Gravitational-Wave Observatory, reported that researches had detected gravitational waves that were created by the merging of two black holes. This confirmed the prediction of Einstein over 100 years prior. A common analogy to help understand gravitational waves used, is of a child bouncing on a trampoline. But consider the following statement discovered centuries earlier:

By the heaven which is characterized by its bounce

Not only is this statement simply addressing a phenomenon, but an oath is being taken to impress upon the reader the truthfulness of it. Whilst everyone knows of gravity, what isn't as commonly known are microgravity. What also isn't commonly associated are astronauts and swimmers. But in order to be an astronaut – you also have to be a very good swimmer. The reason is given by NASA who state that "floating in space is a lot like floating in water". This is also why NASA have built the Neutral Buoyancy Laboratory, to help astronauts train for space walks by simulating the environment in their purpose-built pool. The purpose of the pool is also corroborated by the following statement:

And it is He who created the night and the day, and the sun and the moon; all in an orbit are *swimming*

Gravitational Anchors. This term was coined by Physicist Robert Tuttle in 2012. To help illustrate this concept. Imagine a buoy on the sea. The buoy is connected by a chain to an anchor that is holding it in place. If the anchor is moved, the chain causes the buoy to move too. This in turn sets off waves in the water. If the analogy is translated to gravitational language, then the anchor is the source mass, the chain is the gravitational force and the buoy is the target mass. The waves caused by the buoy's movement, and induced by the motion of the anchor, travel in the water and are analogous to gravitational waves. For the earth, the sun is the gravitational anchor of the solar system. The black hole which is found in every galaxy – including our own – acts as a gravitational anchor for the galaxies, and this is what holds them together in space. All this proves that the earth is anchored in space. As this is known only very recently, then how can the following statement have been made centuries earlier?

> He Made in the earth what anchors it from high above

Big Bang Theory

The two scientists that had proposed this theory were Georges Lemaitre and Alexander Friedmann in the **1920s**. It was based upon the works of Albert Einstein who suggested that the universe was expanding. In 1929 it was Edwin Hubble who observed that all the clusters of galaxies appeared to be moving away from other clusters. The reason why this theory is accepted is because even until now we can see the universe continue to expand. As you can see, this explanation had only been found around the 20th century. This being the case, how then can the following statement be possibly made centuries before?

> And it is We who have built the heaven with [our creative] power; and verily, it is We who are steadily <u>expanding</u> it

> I did not make them witness to the creation of the heavens and the earth or to the creation themselves

The universe began from a concentration of extremely high density of energy and gravitational waves. These are also the same waves from black holes in space, which also affect Earth. Consider also the following:

> In four days, He made in the Earth what anchors from high above it and put *baraka* in it. And equally measured out sustenance (of its dwellers) for those who ask

If we break down this statement, 'from high above it' is referring to the gravitational waves that are coming back from those colliding black holes. These are acting as gravitational anchors that are anchoring the Earth, as well as anchoring things to the Earth. A definition of *baraka* is to grow and increase in volume above and beyond expectations. What this is alluding to is the fact that Earth had started out small but continued to grow and expand. Anti-matter disobeys energy and momentum conservation, symmetric charge conjugation and parity.

Humans made from Stardust

In science lessons at school we learnt about the water cycle: how water evaporates and turns into rain. The question arises: how did water form on our planet? Carl Sagan asserts that this arrival was when stars first appear. Inside stars there are a number of burning stages, that transform the star into an online-like shell structure. These produce heavier elements at different stages. These include: oxygen, carbon, chromium and copper to name a few. The formation of these elements causes the star to expand even more. The last element which is produced and sits at the centre is iron. This element stops the expansion process. The star then becomes a gas, and the size of the iron core is slightly smaller then Earth. The end of a super-massive star results in an explosion which is called the supernova. After the explosion what is left is the core which is called the neutron star or the pulsar. This has unique characteristics: it is very dense; it punches holes in stellar disks. In NASA's Chandra X-Ray observatory, scientists have observed a pulsar punching a hole in a disk of gas around it's companion star. This information was published on their website on July 23rd **2015.** Supernova scatter the elements that surround the core of the star, that are made around the nuclear fusion by the star out in the cosmos. These are the same elements that make up other stars, planets and everything else on Earth – including our bodies.

In her research published in 2014, In astrophysics journal, Goranka Bilalbegović and her research team, described exploding stars as cosmic cement mixers in order to explain the presence of building materials in space. They recorded that when very massive stars die, they explode and litter space with a variety of elements. All the ingredients of cement have been found in such steller remnants. Meaning that stardust is not just dust, it is cement like dust which solidifies after being mixed with water. The oxygen synthesised in stars via a nuclear reaction is dispersed with the stardust through the cosmos, and combines for the first time with the hydrogen to form water: H2O. Water was formed and emerged for the first time from between the cement like dust, and the remnant star core which is more rigid then steel: the pulsar. This means that the dust grains that float through the solar system contain tiny pockets of water, and water has been found trapped inside real stardust. Scientists believe that this stardust rained down continuously on young planet Earth, and brought with it the organic material needed for the eventual origins of life. The discovery of water in stardust suggests that the continuous stardust falls have acted as a continuous rainfall which brought water to our molten planet. Considering that everything stated here has only been discovered a couple of years ago, how then could the following statement become known centuries earlier?

By Space, and by the Pulsar. Have you understood what the Pulsar is? It is the star, which punches holes. So humans must consider from what they have been created from. They have been created from water, ejected. That is emerging from between the Steel and Cement like dust.

Have you not seen that God has sent down water down from the Heaven to make fruits of various colours, mountains of various colours like white, red and intense black, and also to make the various colours of people, moving creatures and grazing livestock.

In 2004 NASA sent a spacecraft called Stardust on a mission to collect cosmic dust from a comet to be brought back to Earth for analysis. Andrew Westphal, a planetary scientist and the study lead of this mission said "By analysing interstellar dust, we can understand our own origins". The analysis found that Stardust is sticky (covered in organic matter) as stated by Dr William Reville. It is soft (amorphous) as stated by Dr Francisca Kemper. It is dark [tar-like] as stated by Dr

Franz R. Krueger. Considering that these adjectives have only very recently been used by these Scientists, how then could the following statements been found in a Codex centuries earlier?

And of His signs is that He created you from dust

Indeed, We created humans from sticky clay

And indeed, we created Humans from potter's clay of altered black smooth mud

Speed of Light

According to Wikipedia, Ole Rømer first demonstrated in 1676 that light has a speed limit and is not universally instantaneous. Great minds such as James Clark Maxwell and Albert Einstein further refined the speed limit. It was not until **1975** that the scientific world had come to a consensus that the speed of light travels at 186282.397 miles per second. The most important thing to note is that the world is not disputing that this could have been known earlier. However, consider the following statement found in the codex centuries earlier:

He arranges [each] matter from the heaven to the earth; then it will ascend to Him in a day, the extent of which is a thousand years of those which you count

[The people] the following statement was addressed to measured time according to the lunar calendar. When it says the distance travelled in one day is equivalent to a thousand years [12,000 lunar orbits] that is equivalent to the established speed of light. What follows is a detailed breakdown of the calculation, but feel free to skip to the next chapter if you accept the given formula.

In order to calculate the speed of light, the frames of reference need to be defined. In terms of the 'local inertial frames' it is 299792.458 km/sec. Inertial means to travel in a straight line, but Earth is orbiting the sun. In order to make a comparison between the nominal speed of light to 12000 Lunar Orbits / Earth Day, in relation to the gravitational field of the sun we get a 11% difference. If the geocentric frame is inertial then there is a **0% difference.** We now calculate the lunar orbit of an inertial frame starting from the measured non-inertial frame. When there is a non-rotating local frame in relation to the sun, the moon speeds up as it approaches the sun but then slows down the further away it is from the sun. In relation to the stars, how the moon speeds up and slows down is relational to the forward motion within the orbit, with the same angle the Earth orbits the sun. This frame has a rotational force around Earth. This in turn means that the lunar orbit is affected by the torque force that is around Earth. When the distance to the sun approaches infinity, the effect of the twisting motion decreases. When the energy that was gained from the twisting motion is removed, we can then calculate the total energy and in turn, the length of the lunar orbit that is outside the gravitational fields. As the Earth-Moon system leaves the solar system, the geocentric frame travels in a straight line and 12,000 Lunar Orbits / Earth Day becomes equivalent to the speed of light. The difference in the local inertial frame is 0.01%.

The statement was addressed to a people that measured distance according to the time they had walked somewhere. For example, if they said that a town was three days away, then this meant that this is how long it would take to walk to there. The statement gives a specific time: 1000 years. By using the lunar calendar, they knew that there are 12 lunar months in a year. In order to measure the speed of light we will make a comparison to a local inertial frame. 12,000 Lunar Orbits / Earth Day when the geocentric frame is inertial, then comparing this to 299792.458 km/sec. As an analogy, when you spin your loose clothes fly outwards, and as you slow down, they come closer to the body, until you stop, and the clothes are settled on you. The same principle also applies when the Earth-Moon system either gains or loses kinetic energy: the distance to the moon changes.

Humans have always seen the same half of the moon from earth. This has been true for over 4 billion years. The Earth rotates on its' own axis: the same is also true for the moon. This being the case the moon travels 360° around Earth in relation to the stars. When the moon first formed it was very close to the Earth, which meant that it was able to rotate the Earth every few hours. Now it takes 27 days. As the moon continues to move further away from Earth, its' spin continues to slow down. At a certain distance the Moon will stop spinning completely. As the distance to the sun increases to infinity ø decreases to zero and lunar orbit loses this twist. When we remove the energy gained from this twist, we can calculate the total energy and hence the length of the lunar orbit outside gravitational fields.

Equation:

Distance traveled by angels in one day = 12000 x Length of lunar orbit.

$$\Rightarrow C\, t' = 12000\, L'$$

Where:

- **C** Is the speed of angels, which we intend to calculate and then compare to the known speed of light (no external forces; no acceleration; no deceleration)

- **t'** Is Earth Day outside gravitational fields i.e. time for one rotation of Earth about its axis with respect to stars.

- **L'** Is the length of the lunar orbit outside gravitational fields (no external forces; no acceleration; no deceleration)

Time \ Frame of Reference:

This table below shows the lunar month and Earth day in both sidereal system (with respect to stars) and synodic (with respect to sun):

Period	Synodic (sun)	Sidereal (stars)
Earth day t	24 hours = 86400 sec	23 h 56 min 4.0906 sec = 86164.0906 sec
Lunar Month T	29.53059 synodic days	27.321661 synodic days = 655.71986 hours

Every new moon begins at a different point in orbit. It only returns to the same point after 27.3 days. When it returns to the same point in relation to the stars, the Earth-Moon system moves 26.9° around the sun and not 29.1°

Hence it becomes important to distinguish non-inertial motion:

1. If you are in a spaceship and fire your rockets, then you are not inertial.
2. If you are orbiting the sun, then a gravitational force is accelerating you towards the sun; hence you are not inertial either (even if your tangential speed around the sun remains constant).

You can find the answer in:

'General Relativity', Lewis Ryder, Cambridge University Press (2009).

Page 7: "There are, however, two different types of such [non-inertial] motion; it may for instance be acceleration in a straight line, or circular motion with constant speed. In the first case the magnitude of the velocity vector changes but its direction remains constant, while in the second case the magnitude is constant but the direction changes. In each of these cases the motion is non-inertial, but there is a conceptual distinction to be made."

Inside the gravitational field of the sun 12000 Lunar Orbits/Earth Day make 11% difference with 299792.458 km/sec when compared in a local frame non-rotating with respect to stars (sidereal system); however, this frame is non-inertial. When we calculate 12000 Lunar Obits/Earth Day outside the gravitational field of the sun then this frame would travel in a straight line + would not rotate with respect to stars (becomes inertial). The difference in this local inertial frame **is** 0.01%. Finally from the equivalence principle we know that there is no difference between an experiment in a local inertial frame outside sun's gravity and an experiment in a local inertial frame inside sun's gravity, that is, whenever Earth is inertial you will get the same results as if Earth is outside sun's gravity. We chose to calculate the lunar orbit outside sun's gravity because it is easier to calculate however these two experiments are identical.

You can find the equivalence principle in:

'Einstein's General Theory of Relativity, With Modern Applications in Cosmology', Øyvind Grøn & Sigbjorn Hervik, Springer (2007).

Page 15: This means that an observer in such a freely falling reference frame will say that the particles around him are not acted upon by any forces. **They move with constant velocities**

along straight paths. *In the general theory of relativity such a reference frame is said to be* **inertial**.

Einstein's heuristic reasoning also suggested full equivalence between Galilean frames in regions far from mass distributions, **where there are no gravitational fields**, *and inertial frames falling freely in a gravitational field. Due to this equivalence, the Galilean frames of the special theory of relativity,* **which presupposes a spacetime free of gravitational fields**, *shall hereafter be called* **inertial reference frames**. *In the relativistic literature the implied strong principle of equivalence has often been interpreted to mean the physical equivalence between freely falling frames and* **unaccelerated frames in regions free of gravitational fields**. *This equivalence has a local validity; it is concerned with measurements in the freely falling frames, restricted in duration and spatial extension so that tidal effects cannot be measured.*

When the Earth-Moon system is still inside the solar system the position of the sun relative to Earth with respect to stars changes; this means that the moon has to make more than 360 degrees with respect to stars in order to point to the sun again. However, when the Earth-Moon system exits the solar system the position of the sun relative to Earth with respect to stars remains the same, that is, the moon now only has to make 360 degrees with respect to stars in order to point to the sun again (the synodic periods become equal to the sidereal periods). This means that the lunar month with respect to the sun becomes equal to lunar month with respect to stars and Earth day with respect to the sun becomes equal to Earth day with respect to stars. The 1000 lunar years with respect to sun become equal to 12000 lunar months with respect to stars.

Length of Lunar Orbit

Length of lunar orbit = Velocity x Time (L = V T)

In a local frame non-rotating with respect to stars the velocity of the moon is not a constant. NASA measured the instantaneous velocity of the moon at various points throughout its orbit. These measurements show that the velocity of the moon varies considerably (from 3470 km/hr up to 3873 km/hr); which means that the moon accelerates and decelerates continuously. The average lunar velocity is V_{avg} = 3682.8 km/hr (1.023 km/sec).

    The lunar orbit relative to Earth is a low eccentricity ellipse, however we cannot use the equation for the perimeter of an ellipse. Why? Because Earth lies on the major axis of this ellipse; but since the direction of the axes change with respect to stars then when the moon returns to the same position with respect to stars, this does not mean that it made an exact ellipse (a local frame non-rotating with respect to the ellipse is actually rotating with respect to stars). Most astronomers calculate the length of the lunar orbit in a local frame non-rotating with respect to stars by the following equivalent circle method:

$$L = V T = 2\pi R$$

$$\Rightarrow V = 2\pi R / T$$

However, this velocity is under the influence of the gravitational pull of the sun. We can vectorially calculate the velocity of the moon relative to Earth without the gravitational assistance of the sun and hence the isolated length of the lunar orbit: Displacement is a vector (has magnitude and direction) and from this displacement vector we get the velocity vector (magnitude and direction); and from this velocity vector we get the kinetic energy. If external work is done, we end up with a resultant displacement vector, resultant velocity vector and resultant kinetic energy.

Ocean Friction:

What causes oceanic high tides and low tides? Lunar gravity. The following demonstrates how ocean friction transfers the kinetic energy of Earth's spin to the lunar orbit. As the moon acquires more energy from Earth's spin it does not actually speed up, instead it slows down because it orbits at a higher altitude (R' increases). If Earth were spinning in the opposite direction the reverse process would have happened; the moon would have lost altitude and eventually crashed into Earth. Today's lunar orbit is a very low eccentricity ellipse (very close to a perfect circle) but when the moon first formed it was a very high eccentricity ellipse. The eccentric ellipse back then had a point very close to Earth and another point very far out. When the moon was nearest to Earth inside the ellipse the gravitational forces were stronger, hence more energy was transferred to the moon when it was closer to Earth than when the moon was farther out inside the ellipse. This made the moon recede more when it was closer to Earth than when it was farther out inside the ellipse. This difference in recession rates smoothed out the differences between the closest and the farthest points in the orbit (that is why todays lunar orbit is very close to a perfect circle). For each direction the recession has different magnitude. Since this recession has magnitude and direction then it is a displacement vector (**R'**). However, in every direction this displacement vector is normal to the rotational force around Earth (always at right angles with the rotational force around Earth, 90°). Since the resultant **R** is the sum of two normal displacement vectors then those displacement vectors form a right triangle. We can use trigonometry to solve those displacement vectors:

By definition in a right triangle $\cos\theta$ = **side adjacent / hypotenuse.**

$\Rightarrow$ side adjacent = hypotenuse $\cos\theta$

$\Rightarrow$ **R' = R** $\cos\theta$

We can verify this triangle by the Pythagorean Theorem:

$$(R\sin\theta)^2 + (R\cos\theta)^2 = R^2\sin^2\theta + R^2\cos^2\theta = R^2(\sin^2\theta + \cos^2\theta) = R^2 \quad (1)$$

Hence cos ø is the only solution to this restricted three-body problem.

To learn why there are two tidal bulges, one facing the moon and another one opposite to it.

Inertial Earth-Moon System:

The lunar orbital radius **R** is a function of total energy however, the total orbital energy comes from two sources: **ocean friction** and **gravitational twist** (two sources, not one). As the distance to the sun increases to infinity the lunar orbit loses this twist; but without the energy gained from this twist the orbital radius decreases to R' = Rcosø.

You might ask why can't the moon simply remain at radius R and velocity Vcosø? Well the answer to this is to actually counter intuitive. If you were driving a car and apply the brakes, then the car will decelerate (lose kinetic energy). But if you apply brakes to the moon (lose kinetic energy) the moon will slow down for a while however it will become unstable (too high too slow). The moon that was at the right velocity around Earth now becomes a little bit too slow for Earth's gravitational force (shortage in kinetic energy). To compensate for this imbalance the moon descends to a lower altitude (trades excess potential energy above Earth with kinetic energy). After descending a bit the moon speeds up and returns to equilibrium. So contrary to driving a car, if you apply brakes to the moon it will eventually speed up! This might sound crazy however it is true. In our case the orbital radius decreases to R' = Rcosø and the moon accelerates to. This means that without the energy gained from this twist the orbital radius decreases to R' = Rcosø. Hence when inertial the length of the lunar orbit becomes **L' = 2πR' = 2πRcosø = Lcosø** (i.e. 12000 L' / t' = 12000 Lcosø / t') and the orbital period decreases.

Today when the moon makes 360 degrees around Earth with respect to stars the Earth-moon system moves 26.92952225 degrees around the sun. Hence the lunar orbit's twist angle ø = 26.92952225 degrees. We can calculate ø from the period of one heliocentric revolution of the Earth-moon system (365.2421987 days):

ø = (360 degrees / 365.2421987 synodic days) x 27.32166088 synodic days = 26.92952225 degrees
Similarly, Earth's spin slows down by 6 seconds i.e. the isolated Earth day t' becomes 86170.43114 sec.

Compare:
Now we can check the accuracy of this equation:

C t' 12000 L'

The distance traveled by light in one Earth Day = C t' = 299792.458 km/sec x 86170.43114 sec = 25833245358 km

The distance traveled by angels = 12000 L' = 12000 x 3682.8 km/hr x 655.71986 hr x cos(26.92952225) = 25836303825 km

By dividing the two distances we get the ratio of 1.00011839267.

Difference = 0.01%

Biology

Worker Bees are Female

In William Shakespeare's Henry IV, he describes the worker bees as being male. No one had criticised him by stating that they are *female*. The credit for this discovery goes to Johann Dzierzon in **1845**. The following statements were made in Arabic, and this is important because a lot of the terms used are regarded in the feminine form i.e. describing females. In Arabic the word *Kuli* 'eat' is used in the feminine form. The word *Usluki* 'to follow a path' is also a feminine expression. The word *Batuniha* 'belly' is also a feminine word. Therefore, the statements inform the reader that the worker bees are female. How could anyone do this when it was universally assumed that worker bees were male?

And your Lord inspired to the Bee "Take for yourself among the mountains, houses [i.e. hives] and among the trees, and [in] that which they construct.

Then eat from all the fruits and follow the ways of your Lord laid down [for you]." There emerges from the bellies a drink, varying in colours, in which there is healing for people. Indeed, in that is a sign for a people who give thought.

Hearing before seeing

The earliest research paper was published in **1994**[1] What was determined was the hearing in the foetus begins development after 19 weeks. At 27 weeks the baby's eyes open. In the course of human history, this knowledge is pretty much brand new – that hearing develops before sight. It would be perfectly reasonable to assume it's the other way around, because of how much more reliant humans are

---

1 Development of fetal hearing | ADC Fetal & Neonatal Edition (bmj.com)

with seeing then hearing. Then how could the following statements have been made centuries earlier, correctly placing hearing before seeing?

[For] indeed, We [alone] created humans from a drop of mixed fluids, [in order] to test them, so We made them hear and see.

He is the One Who created for you hearing, sight, and intellect. [Yet] you hardly give any thanks.

High Altitudes Affect Breathing

The higher humans approach the sky, the pressures of oxygen levels decrease. The result is that humans need assistance with breathing by using ventilators. Before this stage humans begin to breath faster, and physical movement becomes more difficult. Also, as a person goes higher they also find that the amount of fluids they retain decreases, as well as becoming hungry quicker due to the increase in metabolism. However, the longer a person stays at that altitude their body is able to adapt to the situation by being able to change/sustain the new breathing levels, as well as the heart rate changing in order to accommodate supplying oxygen to the body accordingly.
    The Bone Marrow handles red blood cells. At higher altitudes more blood cells are produced in order to help oxygenate the body. However, too much cells being produced can lead to the blood thickening and inadvertently leading to clotting. The need to urinate will also increase as the body will try to dissipate the abnormal fluid levels. Sleeping also becomes affected as the brain is subconsciously reminding you to breath. It is difficult to determine how a person becomes affected from altitude sickness. Mountain trekkers can still be affected by it regardless of how many times previously they have walked. People that are known to suffer from hypoxia, which is where they are unable to receive the adequate levels of oxygen in normal circumstances, should particularly be careful if they find themselves in higher altitudes. Thalassemia is also another condition that can have

detrimental affects on a person in higher altitude. This is a blood disorder where it doesn't contain sufficient levels of oxygen.

What happens to air at higher altitudes?

The composition of air (nitrogen 78.09%, oxygen 20.95%) remains the same at all levels. The "partial pressure" of oxygen changes at high altitude. It means the percentage of oxygen present in the air is less, in comparison to the air at the sea level. With the air molecules pressed together, oxygen is more concentrated at lower altitudes. But the molecules in the air are not pressed together at high altitude. The lack of pressure reduces the concentration of oxygen molecules present in the same amount of air we inhale. This condition is often referred to as 'hypoxia'.

How does the change in altitude affect the body?

The human body is likely to respond to a severe change in altitude. Any high-altitude place is an overdrive, so the body may be losing energy even while resting.

- Lungs: The lungs play a crucial role as it acts as an interface between the environment and the body. The body is resilient as it can adapt to any abnormal environment (in this case the process of acclimatization) to overcome the thin air. This causes <u>acute mountain sickness</u> (AMS) which is accompanied by headache, gastrointestinal problems, and sleep disturbance.

- Blood: The levels of haemoglobin [the oxygen-carrying protein] increases. The steady flow of oxygen in the blood ensures the muscles function properly. However, excess haemoglobin in the blood can result in severe clotting.

- Brain: Affects to the brain are rare, but if they do occur the results can be deadly. The initial symptoms of altitude sickness are nausea, vomiting, headache, insomnia, fatigue, and dizziness. The condition gets worse when a person develops High-Altitude Cerebral Oedema (HACE), i.e. the swelling of the brain. HACE can also result in causing deaths.

The lack of oxygen damages the brain cells.

- Kidney: There is an increase in the frequency of urination as the body tries to get rid of excess fluids. This reaction causes Kidney diseases to progress faster at high altitude which can lead to renal hypoxia and renal injury.

- Heart: Patients suffering from cardiac issues should avoid places of thin air as they are more likely to suffer heart attacks. The decrease in oxygen causes the heart to pump faster, increases adrenaline release and pulmonary artery pressures.

Why acclimatization for high altitude treks becomes necessary?

Altitude sickness is caused by climbing high without any rest. The body requires a minimal amount of time to adjust to any given environment. This process of adjusting to the fewer molecules of oxygen is known as acclimatization. The number of oxygen molecules per breath reduces with an increase in altitude. Acclimatization is necessary while trekking as a lack of adjustment can cause life-threatening diseases. It is only when an acute mountain sickness (AMS) turns towards HAPE (High Altitude Pulmonary Oedema) or HACE (High Altitude Cerebral Oedema) that it becomes deadly. Eating and sleeping well, ascending slowly, staying hydrated, and regulating body temperature helps in acclimatization. The body faces less discomfort if it is acclimatized well. It is mostly accepted that the pace of acclimatization is dependent on genetics. The trekking companies keep adequate resting time for the better performance of individuals. As trekking is a recent phenomenon attempted by humans, the whole notion of less pressurised oxygen at higher altitudes would have been unknown before. How then could the following statement have been made centuries earlier?

Whoever God wills to guide, He opens their heart to Submission. But whoever He wills to leave astray, He makes their chest tight and constricted as if they were climbing up into the sky. This is how God dooms those who disbelieve.

Fingerprint

The person who is credited with being the first to discover that every human has a unique fingerprint is Johann Christoph Andreas Mayer in **1788**. What is miraculous is considering the size of fingertips and how many people there are in the world, we still manage to have a print that is unique to the individual. We are so confident in this knowledge that we have used fingerprints as means of identification. As this is the case how then could the following statement have been made centuries earlier unless they knew what the implications were?

> Yes, We are able to to put together [in perfect order] the tips of their fingers

Three Dark Stages

During the development of the human embryo, there are three stages that it goes through:

1. The darkness of the anterior abdominal wall
2. The darkness of the uterine wall
3. The darkness of amniochorionic membrane

Knowledge of embryology was only possible after the development of the microscope and advanced scientific methodologies. This being the case, how could the following statement have been made centuries earlier?

> He makes you, in the wombs of your mothers, in stages, one after another in three veils of darkness

Made from Water

A Scottish Botanist Robert Brown, is credited with categorising the structure of a cell in **1883**. Scientists who came after built on his work to provide more information about the cell: the different elements and their respective functions. One revolutionary discovery had also been made: that around 70% of a cell's composition is *water*. If this discovery had been made only very recently, then how could the following statement be made centuries earlier?

Have those who have disbelieved not considered that the heavens and the earth were a joint entity, and We separated them and made from water every living creature? Then will they not believe?

The Prefrontal Cortex

This is at the front part of the brain, and it has the ability to enable humans to lie. The more a person lies, the more prefrontal white matter develops in that part of the brain. This discovery was made only very recently, and before then humans had no idea which part of the brain could be responsible for lying. So how could centuries before the following statement be made?

A lying, sinning forelock

Bones before Muscles

The first formation of bone is in the jaw and this happens around 41 days. Muscles are then formed around three days after. How can the following statement accurately describe the formation of a baby centuries earlier?

Then We made the sperm-drop into a clinging clot, and We made the clot into a lump [of flesh], and We made [from] the lump, bones, and We covered the bones with

flesh; then We developed them into another creation. So blessed is God, the best of creators.

Gender Determination

What Humans now know is how we reproduce. The male sperm penetrates the female egg, thus begins the development of a human. But knowledge which is even more recent then this is that it is the *male* sperm that determines the gender of the child. The sperm carries the X and Y Chromosome, while the Female egg only carries the XX Chromosome. The German Cytologist Hermann Henkin in **1891** first discovered that by using a light microscope to study the sperm formation in wasps, Henkin noticed that some wasps sperm cells had 12 chromosomes, while others only had 11 chromosomes. Also in his observations of the myosis that lead up to the formation of the sperm cells, he noticed that the 12th chromosome looked and behaved differently from the other 11 chromosomes. This he called the X chromosome. In 1905 Nettie Stevens surveyed multiple beetle species and examined the inheritance patterns of their chromosomes. She noticed unusual pairing of chromosomes that separated to form sperm cells in the male beetles. Comparing the chromosomes between the male and female beetles, she determined that the accessory chromosomes are what determine the gender. Considering how new this information is in light of human history, how then could the following statement be made centuries earlier?

He has created both sexes, male and female, from a watery bit of the semen which has been ejected

Evolution

This theory is credited to Charles Darwin in **1859**. It has two main points. The first is that all life on Earth is connected to each other. If we were to trace the lines of descent of all living creatures, we would see that they all would converge to a single point of origin. Charles Darwin suggests that this point of origin was a small pond. If this became acceptable knowledge only very recently, then how could the following statement had been found centuries earlier?

And We made from water all living things

The second point of the theory is that all life are a product of modifications by natural selection. This is where some traits in an environment are favoured over others. Consider, then, also the following statement:

Your Lord creates whatever He wills and Selects

At the heart of the theory is the idea of copying, and the variations that are produced from this. Does this not verify the following statement?

What is the matter with you that you do not hope for honour with God, when he created you in successive stages

Charles Darwin believed that humans originated in Africa, and this was only corroborated in 1962. Geneticists are able to identify certain genetic sequences or 'markers' in each of us and cross-reference it with a number of ever-growing international databases. Where there's a match, there's likely a common ancestor and genetically speaking, all markers point to Africa. With the mapping of the human genome in 2003, combined with thousands of people around the world submitting their DNA for testing, there's now proof that we all emerged from Africa.

If we were to summarise Human Evolution, we could use the following. The first is considering the Fig. This fruit has been credited with shaping humans. Wild fig trees first grew in Africa over 80 million years ago. Humans have been consuming this throughout their history. By being able to grow all year, this would have been the staple diet of our ancestors. The high energy content would have helped humans to develop larger brains, and human hands would have evolved in order to determine how soft the figs were (for their sweetness). Fig trees are known to have originated in Ethiopia and Yemen.

Dr Guillaume Besnard has researched Olive trees, and determined that the three main branches of wild olive split from the common ancestor at least 1.5 million years ago. Mountains have also been instrumental in shaping human evolution. When humans had first moved to higher altitudes, the biological effects they experienced is what had caused a reconfiguration. An example of this is the adjustments of haemoglobin levels. Professor Brian Villmoare had discovered "The most important transition in human evolution" – and it was discovered in Ethiopia.

Human Evolution is the process that led to the anatomically modern humans. This began with the evolutionary history of primates which appeared 80 million years ago, and leading to the emergence of the Homo Eructus or the 'Upright Man'. This is the first creature to stand fully upright. We first appeared around 1.5 million years ago. If all this information has been agreed upon only very recently, then how could the following summary been made centuries earlier?

By the Fig and the Olive, by the mountain of Sinyin, and by this secure city [Mecca] We have indeed created the human in the best erected shape, then we returned him to the lowest of the low. Except for those who believed and did righteous deed, For them is a reward unending.. So what causes you to deny the day of Recompense? Is not God the most Just of Judges?

The Fig tree was first produced 80 million years ago, the Olive tree was grown 1.5 million years ago. The mountain of Sinyin is found in Ethiopia. Mecca is the city that enabled Humans to reach their fullest potentials. Scientists are now searching the origins of humans in Danakil Ethiopia – which also happens to the be the lowest point on Earth. [2]

---

2   Recommended Reading: Creation and/or Evolution by T.O Shanavas

Miscellaneous

Flying

One of the aspirations of humans was to be able to fly. From Leonardo Di Vinci's flying machine drawings in 1490 to all the subsequent renderings and prototypes by different individuals, humans had never lost the motivation to one day being able to do so. What is credited as the world's first controlled, sustained flight of a powered heavier-than-air aircraft was produced by the Wright brothers, called the 'Wright Flyer' on December 17, **1903**, 4 mi (6 km) south of Kitty Hawk, North Carolina. The brothers were also the first to invent aircraft controls that made fixed-wing powered flight possible. What we take for granted today was still unimaginable only 2 centuries earlier, and so to suggest that humans were capable of this any earlier would [understandably] have been madness. How would it have been possible for the following statements to then have been made centuries earlier?

And you cannot escape Him on earth or in heaven. Nor have you any protector or helper besides God

You will certainly pass from one layer to another

In modern parlance the word 'layer' is substituted with *Atmosphere* and the different levels of atmospheres determine the flight levels. For example, planes fly in the Stratosphere, while astronauts 'walk' in the Exosphere. Who could possibly have had the confidence to say that humans would be able to reach such distances centuries before being able to do so?

Barrier between the Seas

Although humans have been sailing the oceans for centuries, they would have considered the whole ocean as being the same. Only until recently we have discovered this is not the case. Although we cannot ascertain exactly when the following discovery had been made, it cannot be denied that it was recent due to the availability of the equipment needed to substantiate it. The Atlantic Ocean water has characteristics that are different from the Mediterranean Sea. Other seas also have different properties in terms of salinity and density for example. What is even more miraculous is the differences remain even when the two seas are adjacent to each other. What is only now known is that the surface tension caused by the difference in density of their waters is what is preventing them to mix with each other. If this was only known recently, then how could the following statement have been made centuries earlier?

BARRIER BETWEEN SEAS

And it is He who has released the two seas, one fresh and sweet and one salty and bitter, and He placed between them a barrier and prohibiting partition

Iron

What had been assumed by humans is that Iron, like all the other materials found on the planet, are a part of the Earth's formation. While humans have been benefiting from this wonder material for thousands of years, it is only very recently that we have come to know that its' origins is its abundant production during the runaway fusion and explosion from a supernova: the explosion of a star. This debris which scatters the iron into space is what caused it to crash into Earth. This being the case, how then could the following statement have been made centuries earlier?

And We sent down Iron, in which is [material] for mighty war, as well as many benefits for humans.

This codex has 114 chapters. 114/2 = 57. This chapter is called The Iron, and it is placed directly in the centre of the codex. The reason why this is being stressed is because in the middle of the Earth is also a ball of Iron, which has a radius of 760 miles. The temperature is about 5700 Kelvin, which is almost the same as the surface of the sun. Earth is made of layers. The second layer above iron is nickel and iron alloy. This layer is a fluid, but the iron parts are solid and they rotate in the same direction of the planet. However, the speed of the friction is different from the rotation of the planet. This is what causes the magnetic fields. There are many animals that are able to use this magnetic fields in order to help them navigate across the planet. The liquid part of the core is about 1355 miles in thickness. Then we have the third layer which is the Mantle. The layer is not as hot as the second layer but it is much thicker. The layer above the Mantle is the crust of Earth.

The significance of 57 is 19 * 3 (please refer to the subject of 19 in this book). The next is that we are looking at the Iron with an Isotope of 57. The title of the chapter 'The Iron' which in Arabic is Al Hadeed. If we take the geometrical value of these two words, then they are 57. If we break this number down further, then the geometrical value of Hadeed is 26. It just so happens that this is also the Atomic number of Iron. The number of electrons Iron has is 26. If we take the geometric value of Al it is 31. The number of Neutrons Iron has is 31.

Bounce and Crack

What had been discovered only very recently is that when space is stretched out it bounces, while when the earth is stretched out it cracks. Although the darkness of space prevents us from seeing such a phenomenon with our own eyes, in **1916** Einstein's general theory of relativity determined that spatial bodies move in the fabric of space, in a manner similar to how a bowling ball affects how it is placed on a trampoline. In 1789 Iceland's Silfra Fisher was opened. In 1927 the first crack in the Earth was recorded in the Arizona desert. A book from 1962 titled 'Notes on Earth Fishers in Southern Arizona' by Geraldine Robinson and Dennis Peterson published information about cracks in the earth, and they confirmed that the first known crack appearing on earth was in Arizona. Taking all this into consideration, if both these phenomena were discovered only very recently, how could the following statement be revealed centuries earlier?

By the Sky which has the bounce, by the Earth which has the crack. Indeed, what is written inside the book is very serious, and it is not in jest.

Relativity

Einstein proposed that both Space and Time are relative, opposing Newton's theory that both Space and Time are absolute. The only thing constant in Einstein's theory is the speed of light. This will be the same regardless of the observer's uniform motion. There have been many experiments conducted over the last 100 years to prove that it was Einstein's theory that was correct. Many of the systems we now have work on his principles. For example, the Global Positioning System (GPS) and Nuclear Power Plants work on Relativity. Relativity states that there is no absolute frame of reference. Every time you measure an object's velocity, or it's momentum, or how it experiences time – it is always in relation to something else. The speed of light is always constant, regardless of the position is measuring it: if they are stationary or they are travelling while attempting to measure it. Nothing can go faster than the speed of light.

If the speed of light is always the same, then an astronaut orbiting the planet in space is considerably quicker than what they would be experiencing if they were on the planet. How they would be perceiving seconds would be slower than what people on earth would be. This experiencing of slower time is known as time dilation. However, there is also another phenomenon: Length Contraction.

Building on the astronaut example, if you were standing on the planet and you were to see a space shuttle orbiting the planet, it would appear as if the shuttle has been compressed. The reason for this is when it is travelling horizontally, the object's horizontal dimensions is what are contracting and not it's height. The slower the spaceship moves the less contraction force is placed upon it. If it was moving at relative speeds, that are closer to the speed of light, then the levels of contraction become significant.

Professor James Kolata in his book 'Elementary Cosmology from Aristotle's Universe to the Big Bang and Beyond' published in 2015, explains that neither Time nor Space are absolute quantities. He presents an analogy. Imagine a meter stick lying on the floor of a room. With one end against a wall, bright lights are shining from the ceiling and the other wall. Now let us rotate the meter stick while looking at its shadows on the wall and the floor. At first the shadow on the floor is equal to one meter in length, and the length of the shadow on the wall is very small. As we rotate the stick towards the vertical, the shadow on the floor becomes shorter, while the shadow on the wall becomes longer. When the stick is vertical, the shadow on the wall is 1 meter in length. A similar process happens in Relativity. As the speed of an object increases, it's length [in the direction of the motion] becomes shorter, while the intervals become longer. Considering all this information has only been presented very recently, how could the following statement have been made centuries earlier?

Have you not considered your Lord? How He extended the shadow, and if He willed, he could have made it stationary. Then We (God) made the sun it's guide, then We (God) reduced it. Little by Little.

If we analyse these statements further, by stating the shadow being 'stationary' it would be conforming to Newton's theory of the absolute rest frame. By stating the sun being the 'guide', it is referring to the speed of light of the sun being the absolute constant. By stating 'reduced it' it is referring to the contraction. 'Little by Little' referring to Time Dilation. This statement validates Relativity, Length Contraction, Time Dilation and Speed of Light being a constant.

Quantum and Light Coherence

NASA had created Fluorescence Maps. These measure Chlorophyll Fluorescence Light, and the purpose for this is to illustrate a different perspective of the Earth's land plants. The data to produce these maps are collected from a spectrometer upon a Japanese satellite. The data specifically focuses on land vegetation and it covers the entire globe. Chlorophyll Fluorescence offers a more direct window into the inner workings of the photosynthetic machinery of plants from space. For example, if plants are under environmental stress before the outward signs of the leaves turning brown. The light being emitted from plants is outside the spectrum human eyes are able to view, but there is an amazing scene in the movie *Lucy* which shows how it would look. Quantum theory informs us of the coherence of light and that it has a <u>dual</u> nature: it can behave as both particles or waves. When plants receive too much light they glow. This glowing process is derived from the cellular process of absorbing the rays from the sun and in turn emitting fluorescent light. Incredibly, olive oil also emits this light. Olive oil is a product of fruit instead of a seed. It is for this reason that it contains the coloured plant pigment Chlorophyll. If you were to place the oil in a clear glass and shine a green laser on it, the oil will glow red. As well as lasers, ultraviolet lamps can also have the same effect. Before the advent of these specialised equipment, as well as quantum theory explaining this whole process, it would have been impossible for humans to have known about this. Then how could it be that centuries earlier the following statement was discovered?

God is the Light of the Heavens (space) and the Earth. The parable of his Light is as a pillar on which is a lamp. The lamp is in a glass, the glass is like a brilliant planet. Lit from a blessed olive tree, neither eastern nor western, the olive oil whereof would almost glow forth, though fire has not touched it – light upon light!

The statement presents a comparison of a glowing earth with the glow from olive oil using the same methodology: fluorescence. Both are also predicated upon the physical phenomena outlined by quantum mechanics, and both emit a coherent light. Coherence is one of the core concepts of quantum mechanics, and it is strongly related to the ability of light to exhibit interference effects. A light field is known as coherence when there is a fixed phase relationship between the electric field values at different locations or at different times. Coherent light waves are waves with a constant phase difference.

They have the same frequency and wave lengths. When two waves meet, they will interfere and superpose. This is called the quantum principle of superposition. The definition of superposition is 'one upon another.' 'Light upon Light' could therefore be interpreted to the superposition of light waves. Planets are known to have been referred to as Bodies that orbits a star (in our case our Sun) and they do not emit light of their own, the statement categorically demonstrates that not only the Earth and Olive Oil glow, but they do so using the same methods!

Ouzo Effect

Oil and water does not mix. However, there is a process of mixing these two in such a way that it prevents them from separating: emulsification. In **1853** the first emulsifiers such as mono and diglycerides were synthesised by the Frenchman Mars Lamb Vertilo. The fascinating aspect of this process is that while water is clear, and oils come in various hues of yellow, the resulting mixture is always milky white. The word emulsion is derived from Latin meaning 'to milk'. Today's herb liquors are an alcoholic beverage with a high alcoholic strength of 40% or more. Herb liquor as the name suggests, has a string herbal flavour to it and more importantly is transparent. Amazing, when added to water it always changes to milky white. The herb blends that are used to provide this may include anise star, peppermint, camphor, cinnamon, ginger and many others. The production of herbal liquor begins with the distillation of high strength alcohol at 96% and then herbs are added to it. Through the distillation process the oils that are produced become invisible and soluble, which is why herb liquors are transparent. Using an alcohol such as wine or beer for distillation won't work because the alcohol strength is too low. In 1996 it was demonstrated that trying to produce either wine or beer with over 15% alcohol strength wasn't possible using traditional fermentation processes. This is because won't produce over the 15% limit.

Alcohol distillation did not begin until the 12th century, in the school of Silerno in Italy. Fractional distillation was developed in the 13th century by Tadeo Aldorotti. It was not until the 15th century that alcohol distillation was used to make significant quantities of alcoholic beverages. In summary, long before the discovery of alcohol distillation, emulsification and herb liquors the following statements were made centuries earlier:

And they (the people who won paradise) shall be made to drink therein a cup of wine, the admixture of which is ginger

Surely the Righteous shall drink of a cup [of wine] the admixture of which is camphor

A cup of wine shall be made to go around them, from water running out of springs. White, delicious to the drinkers.

An Undiscovered Scientific Codex

Number 19

Over it is 19

The people that were privy to this codex were trying to figure out what this statement meant centuries ago. Throughout the ages they presented their hypothesis, but it is only very recently that we can present something with a mathematical basis. The first question that would arise is why 19? Why not 18 or any arbitrary number? Numbers represent value, and therefore 1 is the first number. Not 0 because that doesn't have any value. The last number of value is 9. Just like how the alphabet from A to Z can be used to produce every known word, numbers can be used to explain how everything *in the universe* functions. Now that we understand that we can use mathematics to explain formulas, the number 19 itself can be considered a marker i.e. to represent something. By stating 'Over it...' is letting us know that there must be things in the universe that has this marker. Maths is also amazing because it is understood universally among humans regardless of the language they speak. Even more amazing then this, however, is that regardless of the written script people use – the numbers 1 and 9 <u>are the same.</u> The following phenomenon are scientifically proven:

- The Sun, the Moon and the Earth become aligned in the same relative position every 19 years
- Hayley's comet visits our solar system around every 76 years. 76 = 4 * 19
- Duration of pregnancy in human beings, from the time of fertilization until birth is about 266 days or 38 weeks.
    - 266 = 19 * 14
    - 38 = 19 * 2
- What is being suggested is that the number 19 can be considered a signature. In order for humans to write a signature they need to use a hand. It just happens that the human hand has 19 bones.
- The wrist has 8 bones. 19 is the 8th prime number.
- The light from the sun reaches Earth in 499 seconds using the scientific average. 499 seconds is 8 minutes and 19 seconds

Mathematics

The codex has a statement composed of 19 letters. The numerical values of the letters are illustrated below:

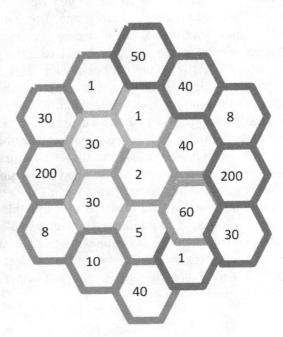

By adding their values together, we can determine the values of each word, and then the total values of all words. So for the first example, we will add the values of the peach hexagons together, and the sums for the remaining examples will also be shown below:

Total numerical value of all words: 786.

The statement has 4 words and 19 letters. By including the numerical number 786 we can do the following:

419786 / 19 = 22094 → No Remainders → All of the following answers will have no remainders

If we were to rewrite the sequence number of each word above and afterwards, the number of letters in each word – the whole sequence is an 8 digit number divisible by 19

13243646 / 19 = 19 x 36686

If we add the number of letters in each word, the number of letters in the next word we will get a 10 digit number:

1327313419 / 19 = 69858601

If you add the numerical value of each word and the number of letters in this word we will get a 15 digit number divisible by 19

110527033354295 / 19 = 5817212281805

2	1	1	1
+ 60	+ 30	+ 30	+ 30
+ 40	+ 30	+ 200	+ 200
+ 102	+ 5	+ 8	+ 8
+ 3	+ 66	+ 40	+ 10
	+ 4	+ 50	+ 40
		+ 329	+ 289
		+ 6	+ 6
1105	270	3335	4295

If we add the numerical value of each word, and the numerical value of the next word, we will get a 16 digit number divisible by 19

1102216834974786 / 19 = 19 x 58011412367094

2	1	1	1
+ 60	+ 30	+ 30	+ 30
+ 40	+ 30	+ 200	+ 200
102	+ 5	+ 8	+ 8
	66	+ 40	+ 10
	+ 102	+ 50	+ 40
		329	289
		+ 66	+ 329
		+ 102	+ 66
			+102
1102	2168	3497	4786

If we write the number of letters in the word, and then the numerical value of each letter in this word separately. The numerical value of each word, we will get a 48 digit number divisible by 19

310226040466130305632913020084050628913020081040 / 19 =

16327686340322647664890158951792138363843162160

2	1	1	1
+ 60	+ 30	+ 30	+ 30
+ 40	+ 30	+ 200	+ 200
	+ 5	+ 8	+ 8
		+ 40	+ 10
		+ 50	+40

3102 26040466 1303056329 130200840506289 13020081040

If we write after the numerical value of each letter it's sequence number, from 1 to 19, we will get a 62 digit number divisible by 19

2160240314305306571830920010811401250131143015200168171 0184019 / 19 =

113696858647647714306890526884810592112165421852640430 0536001

The 19 letters that this whole mathematical precision is derived from are:

بسم الله الرحمن الرحيم

The approximate translation is "In the name of God, the Entirely Merciful, the Specifically Merciful".

The End

Humans often wonder how the world will end. Would we have the same fate as the dinosaurs where a meteor had wiped them out? Or would there be some other form of natural disaster such as the flood that is described in the Old Testament? Dr Benjamin Allanach, a theoretical Physicist at the University of Cambridge states that 'There is a particular particle that exists in the universe which is called the Higgs Boson. The observed 126 GeV Higgs Boson mass seems to imply that that the universe does not exist in the lowest possible energy state, but is positioned in a slightly unusual place".

If the Higgs Boson mass were really 127 GeV and the top mass were a little lower then it's most likely value, then the universe would be completely stable, and the vacuum would be in the true minimum. In the simplest assumptions, the measured mass of Higgs could mean that the universe is unstable and destined to fall apart. Now that we know how precarious the universe is and could collapse at any moment, why hasn't it done so already?

In **1960** the British Scientist Peter Higgs predicted the existence of tiny particles all around us. These particles are responsible for giving objects mass, and also explain why particles behave the way they do. On March 14th 2013 the CERN observatory confirmed that they had found this particle. Professor David Miller provided an analogy to help explain the Higgs Field "Imagine a cocktail party of political party workers who were uniformly distributed across the floor, all talking to their nearest neighbours. The ex Prime Minister walks in and many of the workers will run towards her and form a cluster around her. As she moves other workers will run towards her, and the initial group of people will leave her to make way for them. This movement of running towards and away from her is what causes her to acquire mass". How do we explain the Higgs Particle?

Imagine the party workers are uniformly spread in a room. Now a rumour spreads among them that. It starts from someone at the door spreading it to the people near the door, and they cluster around him. They then inform the workers who are around them, and after doing so return to their original positions. This process continues as the rumour spreads from one end of the room to the other.

The Higgs Boson particle is what gives shape, size and mass to everything in existence. The particle that was found at CERN had a mass of 126 GeV and this made the Physicists nervous, because they believed that it should be 127. This discrepancy is what should cause the universe to collapse. As we now know how the universe will inevitably end, how is it that the following statement could have been made centuries earlier?

It is the day when people will be like scattered moths, and the mountains shall be as loosened wool

So I [God] swear by the clusters you can't see, as soon as they run, they disappear

In the original Arabic there are three words that are used: Khonas and Al Jawary al Khonas. Khonas refers to something invisible, and Al Jawary Al Khonas means something that is running. Al Khonas refers to something that disappears and returns back to it's original spot. If we put the three words together then what we're essentially referring to is the invisible clusters that run and disappear; almost as soon as they start running, they stop and disappear. So what are the clusters this statement is referring to? They are the Higgs Boson Particles.

The universe will end due to mass loss.

And We shall show them signs
In the universe
And within themselves
Until it becomes clear to them
That this is the Truth

In this book what I have attempted to demonstrate are just some of the statements derived from a codex. I wanted to appeal to the type of person who considers themselves rational; and this state of being is achieved through the objectivity of the various sciences. By presenting a hypothesis, then testing it through experimentation, removing all biases via prejudices or any preconceived notions, and only caring about the results through empirical evidences, can the true conclusion be reached.

If what you have read sounds familiar, it's because it is similar to the Scientific Method that is taught in schools. Have you ever wondered from where this was derived? Consider the following:

And [remember] when Abraham said "My Lord! Show me how the dead come back to life". God said to him "Do you not have faith in this?" Abraham replied "I do have faith, but I just want to assure my heart". God said "Then bring four birds, train them to come to you. Then cut them into pieces and place them on different hilltops. Then call them back – they will fly to you in haste, and so you will know that God is Almighty, All Wise".

- My Lord show me... This is the <u>Aim</u>
- Cut them into pieces... For this you need <u>Apparatus</u>
- Train them; Call them back... This is the <u>Method</u>
- Fly to you... This is the <u>Result</u>
- God is... This is the <u>Conclusion</u>

Having reached this far into the book, I must now make a confession which I'm certain many of you may have suspected. The reason why I was not completely honest since the beginning is because of the biases you may have towards Religion. I wanted to appeal to the person interested in Science, and help them realise that the two are not mutually exclusive, but it is the very truth of Science that validates the truthfulness of Faith. I hope to ignite in you, the interest of further researching the Codex, and hopefully find other information that is true to you.

After presenting each Scientific fact, I revealed statements from the Codex that were made centuries prior – they [in fact] were made over **1400** years earlier.

This Codex is the Qur'aan – The Word of God

Citations

Below are all the statements – Verses – that I had used throughout this book

32:5	35:27-28
	30:20
21:32	37:11
16:68	15:26
16:69	76:17
76:2	76:5
23:78	37:45-46
29:22	21:30
84:19	28:68
55:37	71:13-14
25:53	95:1-8
86:11-14	53:45-46
25:45-46	57:25
35:24	74:30
11:86	39:6
21:33	23:14
41:10	21:30
51:47	96:16
18:51	101:4-5
41:10	81:15-16
86:1-8	2:260

References

All the scientific findings are taken from Wikipedia and YouTube

Bonus

There used to be a TV show called *Movie Watch* where each week they would get four members of the public to come in and review movies of the week. Each guest would give marks out of 10. There were quite a few films that got the full 40, but there was one film in which I got convinced into watching: because of the sincerity each reviewer gave as to why it got the full 10 from them.

A few days after I got a call from a friend, who sounded very excited and wanted me and another friend to watch a movie that was about "friendship". At that time, the closest cinema was two train stops away, and we made arrangements to watch it. When we did and came out of the cinema after, it was dark. We just looked at each other and silently agreed to walk home - an hours' journey. Throughout the walk we hardly said a word to each other, but were sharing the same thoughts with regards to the film. That film was *Good Will Hunting*.

Although there are films that I have enjoyed more, I can truly say that this is the only film that I can see a million times - and then watch it again like it's the first time. The film rightfully got the recognition it deserved by winning Oscars for Best Original Screen play and Best Supporting Actor. Apart from this, however, there are no materials available to help people appreciate what this work actually is: there are no features on the DVD; no materials produced by others about the film, nothing other then the film itself or the original screenplay for the film?

My only hope is that the reader would appreciate why the film has moved me enough to write this work. I loved the scene from Rambo where the Colonel is telling him a story about an Artist who created the most beautiful sculpture, and the people asked him how he did that. He replied "I didn't do anything: the sculpture was already there; I just took away the rough edges..." What I'm writing here is nothing new; I'm just trying to help people see the same things I witnessed inherent in the film.

Everybody has a characteristic that is unique to them, but this uniqueness can be a marker whereby others can identify them. Danny Elfman is known for composing certain types of themes, which seem to work particularly well with the Super Hero genre. What is surprising to me is how Mr Elfmans' score for this film is so different from his usual work—but more importantly – how perfectly it sets the context. The piano elements in particular give the impression of someone who is alone, and trying to hide the sadness deep within.

Along with the opening score, we get to see glimpses of Will. The effect of the shot is likened to an insects' perspective, giving the impression that we are in the presence of someone huge. What compounds this feeling is the dark yellow overtone of the shot. The particular hue is known for stimulating ill temper amongst adults, and is also difficult to perceive. The opening scene has perfectly encapsulated the character of Will Hunting without him even uttering a word!

A notion that is prevalent throughout the film is dichotomy. When Ben Afflecks' character Chucky comes to collect him, we witness the poor neighbourhood and the car: we can deduce the social standing of the lead characters. We then witness [gracefully] other parts of the city that subliminally state a truism associated with all cities of the world: there are rich parts, and there are poor parts, *and never the twain shall meet.*

After viewing an evidently prestigious University, we are in an auditorium full of students. What surprised us are the contents on the Blackboard. Human nature tends to drift away from anything deemed difficult, yet here not only are there a lot of people, but they actually seem to be enjoying themselves: they laugh at the Professors' jokes; they are in rapt attention; in essence they are the perfect students. An interesting characteristic of the Professor also becomes immediately apparent: he informs these students that he has a problem and then challenges them to solve it. The real question that should arise is - why? Why would he feel the need to do this when they have already proven themselves by working hard in their lives to get to the point of being in that University? To continue to succeed in their current institution they will need to maintain the same level of commitment... but Professor Lam beau is looking for something more.

Will comes out as a janitor, and this explains his living conditions. At this stage we can presume that maybe he is a student but has not come from an affluent family like most of the students who attend that University: he needs to do what it takes to complete his Degree? The expression on his face when looking at the problem on the Blackboard is of intrigue, but there is no follow up. The next scene jumps straight to a bar, where Will is hanging out with his friends. Because of their age group, it can be assumed that they too are also students in the same circumstances, which is why Will appears comfortable hanging out with them. When he makes his excuses to leave them, and we see him working out the formula at home: Will is a more dedicated student then his friends. Another presumption can be made here in that he is a Maths student.

When Will is doing his cleaning, he makes sure that everyone has left before he starts writing on the Blackboard. This is an extremely interesting characteristic because most would assume that if they knew something like that, they would want to proclaim it to the whole world and thus be acknowledged for their intelligence. Instead he is playing Baseball with his [now apparent to us] best friend, and doesn't mention working out the Maths problem. We are still under the assumption that Chuck is a student because he suggests going to a "Harvard bar", and so perhaps out of respect for their friendship, Will does not disclose his Academic achievements to him.

The following scene shows a class reunion with the standard features. What further substantiates the dedication of the Professors' students is when they approach him - on a "Saturday!". However, the reason soon becomes apparent when they inform him of someone already proving the theorem. He finds it hard to believe because he goes to the board where other students are already checking it out. When the Professor realises it is right, he asks some of them there if they had done it. Why he assumes they have is because he addresses them by their names: they must be special for him to know them. When they deny it, it can be assumed that there is an underlying modesty in these types of students, which is why Will doesn't tell anyone it was him?

When Will is hanging out with his friends [again] it presents a strange question: he always seems to be drinking and generally hanging out, when does he find the time to study? Also his friends don't seem to have any interest in studying. If anyone is attending that type of University, then they would presumably be with other 'book worms' who evidently spend all their time around their studies. Instead they just seem to drive around and argue over the most trivial of things like food. When they see the other gang and make a decision to actually fight them, then this is shocking to say the least! That is one of the forms for gross misconduct, which would result in the expulsion from the University. What is even more hilarious is the only one from them who is trying to talk the rest out from fighting, is doing so for the reason that they are now eating and it would ruin their appetite!

When will approaches Chiromine, he immediately attacks him. This compounds the shock factor to the audience because you would assume that as he is an 'intelligent student' he would at least try and talk to him first; make the guy realise what he had done to Will and at the very least try to gain an apology. Even though there were an equal number in each group, avoiding confrontation would have seemed the best solution, especially when Will had just witnessed them abusing a woman on the street and clearly were not reasonable people.

The ferocity of Wills' attack was shocking because even though clearly Will had hurt him, he continued; even when the Police had arrived, Will's only intention was to inflict as much punishment as he could. A person can only enter that state when in deep anger, and Will must have endured long physical abuse from him? When the Police pull him away from his victim, they [understandably] deal with Will quite physically. As Will is already in a state of adrenalin, he perceives that they are attacking him and subsequently lashes out at them. When they forcefully arrest him, he finally realises his mistake, and we assume that he is thinking that he may now have jeopardised his education.

When Professor Lambau enters his unusually full class, the assumption of Wills' regret in the previous scene could be transposed on to here: that the reason why he hadn't disclosed his work to his friends was because he wanted to make sure that no one would know until this moment: he would present himself amongst all his peers as the "mysterious Maths Magician". However, this event exposed the competitive nature of the Professor. Although he was disappointed that no one had acknowledged credit for the work, he proceeds to the tell class that he along with his colleagues have already written up a more difficult theorem which took "two years to prove!". If Will was there to acknowledge his work, would the Professor simply have congratulated him and after giving him his promised accolades, left Will on his merry way?

When Will is released on bail, we see him working on the new theorem. The Professor and his colleague see him but do not recognise him. Why? Would not the first thing Will had done was immediately approach him and tell him it was he worked out the previous theory? After explaining his circumstances with the arrest (sugar coating of course!) the Professor would [presuming] have arranged for all the Maths students to converge again and give Will the reception he deserved? Instead the Professor confronts Will and thinks that a Janitor is messing around with peoples' work, and therefore is upset about that. When he tries to ascertain his identity, Will walks away and goes through the one of many doors. The Professor decides to not pursue the matter, but is shocked to return and see what is on the Blackboard: the Janitor wasn't messing around but now seems to be the one!

What does Will do when he realises he's been made? He quits! What's going on? His friends ask that when they head towards the bar but he deflects the question. He uses the same excuse his friend used, which inadvertently provides a silent gesture to his friends to not pursue the topic. This they don't mind because by then they arrive at the Bar. At this stage we still think they are all students because they each greet the doorman as they enter: they must be regulars at the Harvard bar, ergo Harvard students. When Chucky casually asks for a drink and sees some "Honeys... need to make moves" the notion of him being a student is further emphasised. He's charmingly acknowledged by the two ladies, and proceeds to attain familiar ground by enquiring if they too are students - this could then be used to lead on to other things to talk about. They humour him by asking which class he saw them in, and when he replies "History" and not get a denial from them, he no longer delves into it and tries to just have a conversation with them.

However, it's not long before he's interrupted by another student. When he presses Chucky about the course, it becomes evident that he's not interested in it, but is trying to show him up. Skyla defends Chucky by telling the guy to "go away" she's not impressed with him, and Will instantly likes her because she seems to be on the level and not stuck up like the other 'intelligent' students. When he makes a joke at Chuckys' expense, it is no longer funny and when he says "is there a problem here?" it's a euphemism for "that's enough; leave it." However, instead of doing so they guy proceeds into a discourse to try and make himself look intelligent and Chucky ignorant. Before Chucky loses his composure and ends up doing something he regrets, Will steps in delivers one of the first great monologues:

> *Hold on... of course that's your contention, you're a first year grad' student who's just*

finished reading maxium Historian, Pete Garrison;

you're gonna be convinced with that until next month when you read James Lennon and talking about how the Economies of Virginia and Pennsylvania were entrepreneurial and Capitalist

way back in 1740. That's gonna last until next year when you're gonna be here regurgitating Gordan Wood; talking about, you know, the pre revolutionary utopia and the capital forming effects of military mobilisation

As a matter of fact I won't because Woods drastically underestimates the impact of...

(cuts in)... Woods drastically underestimates the impact of social distinctions predicated upon wealth, especially inherited wealth? You got that from Vitcus, Organization and Country page 98 right?

Yeah I read that too. Were you gonna plagiarize the whole thing for us? Do you have any thought... of you own on this matter? Is that your thing? you come into a bar after reading some obscure

passage and pawn it off as your own so you can impress some girls and embarrass my friend? The sad fact about a guy like you is that in 50 years time you are gonna do some *thinking of your own and you are gonna come up with two certainties: one; don't do that, and two: you blew 150 grand on a f*&^ing education you could have got for a dollar fifty at a public library!*

After this literal onslaught, the guy can only come back with a feeble hypothetical scenario where he describes the value of a Degree by being in a position to take his family to a skiing trip and Will working in a Diner serving them fries. Instead of being insulted by that, Will casually laughs it off; even acknowledging the possibility, but to him it's better being that then "unoriginal". To stop this debate from continuing further, Will asks him if he still has problems then they could discuss it outside - that is a statement a person of any level of intellectual capacity would be able to understand!

To show that Will is not just another variation of that character; instead of starting to chat up Skyla, he leaves with his gang: a victorious Gladiator. Later when the two ladies are leaving the bar, Skyla leaves her friend and Wills' friend comically excuses himself, she scolds Will for being "an idiot!". Even he's surprised because he thought that he acted like a gentleman by not gloating in his victory; after the extrovert situation he goes the opposite direction by trying to be discreet for the remainder of the time at the bar. She's been smitten by Will and was hoping that he would have spoken to her, in a sense trying to say that by doing so, he had done what he did because he wanted to defend her from the guy as well as his friend. She concedes to Will that the reason why the guy approached is because she knew he was in the same class as her, and thought that he got an opportunity to try and impress her. Will being perceptive also concedes to having worked that one out. Skyla shows an attribute of hers by immediately giving her number to Will: someone she barely knows. She seems like a person who doesn't like wasting time, and respects anyone who is direct. After suggesting coffee, Will subconsciously demonstrates his intelligence further by providing a simile to that particular ritual: the alternative being to chew caramel - "arbitrary" -*adjective 'subject to individual will or judgment without restriction; contingent solely upon one's discretion: an arbitrary decision.'* Be honest: after watching the film you also looked the word up!

You would think that after being shown up for the fraud he is; the guy who tried to show up Chucky would have gone home and think about what he had done, and maybe become humble in the process. Instead, as Will and friends are leaving the bar, they see him in a Diner attempting to chat up another woman! This time Will doesn't hold back and bangs on the window: ensuring he gets the guys' [as well as everyone else there] attention. After Will asks whether the guy likes apples, Will shows that he got her number: apple being a metaphor for Skyla - Will knows that the guy liked her, and by making him indirectly concede that he liked apples/her - Will tells him (and hopefully the woman the guy was attempting to chat up) that 'I got what you like - now what you gonna do about it!' The stunned look on the guys' face and the blatant laughing of Wills' friends at him absolutely reduced the guy to the lonely person he is.

After realising who Will Hunting is, we see Professor Lambau trying to track him down. After entering a shop/office, he enquires if they have a student working for him. He's surprised by the demeanour in which he's being addressed; this shows that he spends little time in the 'real world' and when he needs to mingle with others, they are of similar status. Even when his colleague discloses to them that he Professor Lambau, the guy immediately jokes that his colleague is also a Professor. Who says the working class has no wit? When the Professor is informed of Will attaining the job through his P.O - he is further troubled into finding out the answer to his question: who is Will Hunting?

The Professor enters the courtroom and witnesses for the first time Will in action. He's impressed with the manner in which Will is able to defend himself in court - especially when he puts the opposing Lawyer in his place. However, the court steps in and when he lists Wills' past crimes, the Professor realizes that Will definitely has issues. It is strange that when facing prison, Will then decides to call Skyla and ask her out. But then the real reason behind the call is so that he hopes she will be able to find a way of getting him out of there. I love the lines when the Professor formally introduced himself:

"I'm Gerard Lambau... the Professor you told to go f&^% himself"
"well what the f&^* do you want?!"

Will is surprised when he hears that the Professor has arranged for his release. Of course he then isn't when informed of conditions associated with the release. Will doesn't mind doing the maths, but even the Professor understands the idea of seeing a Shrink is humouring. As we previously saw that Will doesn't want to be in prison, he still does not directly disclose this to the Professor. Will begins his Maths sessions with the Professor, and actually seems to enjoy the collaboration with him. When the Professor pats Wills' head; his associate can't believe this 'punk' who's come from no where is now on such good terms with the Professor. How could he not have the same regard after being with the Professor for so long?

Will reads a book and then it becomes evident that he's done background research on his Psychologist. Instead of acting in a manner that would jeopardize his position with the judge, Will proceeds to confront him about his alleged homosexual tendencies! This compounds the notion that the way Will has been able to survive all this time is by manipulating anyone who he deems as a threat. He's not intimidated by the idea that he's testing someone who's been in the field of Psychology for a long time, and when the Psychologist leaves; Will is there sitting relaxed - a silent job done. We have to appreciate the Professors' openness of trying different avenues of therapy, and we really think that Will is disclosing some part of his history that may explain why he is this current way. However, when he breaks into a song, everyone can't believe what he's done. The Professor knows Will is not co-operating with the therapists, but can't refute Wills' argument of them being the ones leaving. When the Professor suggest that he may know someone who could help, the impression is not because of competence but just realizing that no one else may be able to help Will. The juxtaposition of the outside shot of a campus shows the subtle differences with the one at the beginning of the film: it's very difficult to ascertain the identity of the building, let alone consider it being a prestigious University. The Director helps by having a student walk quickly past the camera to help us make the association.

Robin Williams character is talking to his class, but already we can see that there are no conviction in his words: he's simply rehashing what he needs to. The class are also extremely bored, and he's not impressed with the uninspiring answer to his question. In spite of this he proceeds to give an explanation himself and infuse that with genuine humour that the class acknowledges, and he in turn appreciates that. When Professor Lambau enters, the initial reaction of Sean McGuire is of recognition. When he proceeds to introduce the Professor to the class, we can clearly see him relishing in his adulation. It's only when they step out of the class do we realise that they are associates.

When they converse over dinner, we further realise that they are friends. Why they haven't been in regular contact is because Sean wouldn't hide the disappointment of his 'friend' not attending the funeral of someone. At that moment, Lambau also realises the reality of his decision. The quick scene of Will explaining his 'punishment' to Chucky shows no surprise on his part: it just pretty much confirmed to him that Will can talk himself out of anything. We go back to the Professor trying to convince Shaun to see Will, and provides an amazing story to convince him of Wills' intelligence. What helped was that Shaun too also knew of the story, and the fact that it's the Professor who is arguing Wills' case slowly starts to alter Shauns' decision; shocking still that a genius could be raised up from the same neighbourhood. Shaun could see the desperation in Lambau when he concedes that he visited 5(!) Shrinks before him, and all these reasons finally submits Shaun to [grudgingly] accept.

When Will enters the office, Shaun could see that Will too doesn't want to be there as much as him. Yet he shows his Professionalism by having the other two leave the room: leaving Shaun to focus on Will. Will immediately begins to look around the office, trying to use anything he can as leverage over Shaun. When he questions Shaun about his book collection, Shaun immediately determines that Will must like them in order to have questioned him on it. However, when Shaun tries to pursue the matter, Will starts to deflect. Undeterred, Shaun still questions him and when he concedes that he did read all the books, Will admires that and begins to analyse his literary cannon. While doing so he still tries to find clues that will give him some insight to his new adversary. He quotes books (that I have read!) and scolds Shaun for reading "the wrong f&^&ing books" but when questioned what the right ones are, he implies what is enjoyed... even though they can be wrong in his eyes! Shaun uses any hint he can get from Shaun to try and have a conversation - that will eventually lead to Will trusting him. They both talk about fitness, and before it can be pursued further - Will spots a painting.

Will immediately identifies something within the painting and tells Shaun he thinks "it's a piece of shit". Shaun's not impressed and wants him to expound his opinion because now he too wants to know why it has caught Wills' attention:

> It's just the linear impressionistic makes it a very muddled composition it's also a Windsor Holmer rip off except you got Whitey rowing a boat

> It's Art Monet; I never said I was very good

> That's not what concerns me

> What concerns you?

> It's the colouring

> You know what the real bitch is? It's painted by number

> Is the colour by number? Because I think the colour is fascinating

> Oh really? Why is that?

> I think you're one step away from cutting your f*&^^ing ear off

> Really? You think I should move to the south of France? Change my

name to Vincent?

You ever hear of Sand aining in the storm?

Yeah?

Maybe that means you?

In what way?

Maybe you're in a storm, a big f**%ing storm, the sky is coming down
on your head, the waves are crashing your boat, your oars are about to snap. Your pissing your pants, crying for the harbours. Maybe you're doing what you gotta do, you know maybe that's why you became a Psychologist

Bingo! That's it. Come on let me do my job now, don't start with me

Maybe you married the wrong woman?

Maybe you should watch your mouth! Alright Chief?

That's it isn't it? You married the wrong woman. What happened?
She leave you? Was she banging some other guy?

You ever disrespect my wife I will end you! I will f&^ing end you
you got that Chief?*

Times up!

If Will had simply insulted the painting then Shaun would have just as simply ignored it - Will wanted to provoke a reaction. He therefore criticized the painting professionally, building up to the expose. When Will is giving his interpretation of the picture, Shaun loses his composure and cannot conceal the grief in his voice. When Will gives his conclusion, Shaun thought he managed to escape; but no sooner had he tried to resume the session Will nails it! He turns around and the pure white light illuminating on Wills' face is sub text for Will 'seeing the light - the truth'. To ensure that Shaun will no longer want to see Will again, he tries to pursue the matter and make Shaun feel guilty as hell. Instead he's shocked when Shaun loses all professional composure and threatens him directly. As Will leaves, we might think he too feels a little guilt for having pushed Shaun too far; but that's instantly dismissed when the Professor stands and Will jokes "at ease gentlemen!". Lambau meets Shaun and just from his demeanour can tell pretty much the same thing has happened to him that happened to the others. Lambau is surprised that Shaun [instead] insists on seeing Will again. We have the shot of the paining again, giving us the chance to see it in a new light and to dwell on; but not as long as Shaun who we now see falling into depression. It is assumed that he would think twice about seeing Will again because he's realising the extent to how much he has been ruffled. Will continues to live his life, but now there's Skyla in it. He's having a nice time but when she tells him she'll be going to Stanford after, he jokes about it - but there's warning bells going off in his head.

Will is surprised that it's Shaun who is going to see him again. When they're at the park, Will can't help take further digs because he feels that he got one over on Shaun. Maybe if he tries to be as obnoxious as he possibly could then Shaun will just leave him alone. Shaun confides that not only did he think about what Will said to him, but that he stayed up half the night because of it. However, when Sean tells of his epiphany, Will just wasn't surprised about not being taken seriously by being dismissed as a kid. However again, it's when Sean starts explaining what he meant by that does Will (probably for the first time in his life) begin to start listening.

You've never been out of Austin?

Nope

If I asked you about Art, you'll probably give a skinny on all the
greatest Artists that ever lived. Michaelangelo, you know a lot
about him: life's work; political aspirations; sexual orientation;
him and the Pope, the whole works right? But I bet you can't
tell me what it smells like in the Sisteen chapel? You've never
actually stood there and looked up at that beautiful ceiling - seeing
that. If I asked you about women, you'd probably give me a silver set of your personal favourites. You may have
even been laid a few times. But I bet you can't tell me what it's like
to wake up next to a beautiful woman and feel truly happy.

You're tough kid. If I asked you about war, you'll probably throw
Shakespeare at me right? 'Once more into the breach dear friends'
You've never been near one. You've never held your best friends' head in your lap; watching him gasp his last breaths - looking
at you for help. If I asked you about love, you'd probably quote
me a sonnet. But you've never looked at a woman and felt totally
vulnerable. Knowing that she could just level you with her eyes.
Feeling like God put an Angel on earth just for you. Someone who could rescue you from the depths of Hell.
And you wouldn't know what it's like to be her Angel. To have
that love for her, to be there forever; through anything; through
cancer. And you wouldn't know what it's like to sit up sitting
at a hospital for two months holding her hands, because the
Doctors can see it in your eyes that the term 'visiting hours'
do not apply to you. You don't know about real loss, because
that only occurs when you love something more then yourself, and I doubt you've dared to ever love

someone that much. I look at you; I don't see an intelligent, confident man. I see a cocky, scared shitless kid. But you're a genius Will,

no one can deny that. No one could possibly know the depths

of you. But you presume to know everything about me because

you see a painting of mine and you rip my f^%$ing heart out.

You're an orphan right? (silence). You think I have the first clue

about how you feel? Who you are? How hard your life has been...

because I've read Oliver Twist? Does that encapsulate you?

Personally I couldn't give a shit about all that; I can't learn

anything from you; I can't find out from reading some f&^ing

book. Unless you're prepared to talk about you - and I'm

fascinated Will: I'm in. But you don't wanna do that do you?

You're terrified of what you might say. Your move Chief

Sean knows that Will can pretty much talk about any subject. So he reverse Psychologies him by taking each subject and reiterates each with a single point: it's one thing to know something - it's completely different to actually experience it. Extra weight was given to each point by having Sean speak through experience - the one attribute that can never be attained prematurely. Sean is the first (and probably only) person to perceive what

Will is actually feeling. But he too also realizes that Will is a genius, not from Maths, but simply from the short experience from the Office encounter. Sean knows that Will has never spoken about himself before in his life, and the thought of that possibility is inconceivable. But he makes Will realise that it is the only manner in which he could be understood - and that Will has now found someone who will be able understand what he really means. Will that offer be reciprocated?

It appears that Will is trying to ignore this episode by carrying on with his life, but then he almost tried to talk about it to Skyla when it's soaking wet outside. Will has found a way to deflect any moments that would actually risk his friends asking something remotely serious from him - to joke it off. When Will is back at the office, it would naturally be presumed that Will would just open up the emotional flood gates and go into deep conversations about his life's experiences. Instead he sits there, fully aware of his actions without any considerations to their previous meeting. Sean could have [understandably] been upset about that and just lectured Will into his inappropriate behaviour; but after when he's explaining what just happened to the Professor, he shows his intelligence by accurately reasoning the thought process behind Wills' action - everything must happen on Wills' terms.

We're now in Professor Lambaus' office and he has a friend/lecturer there as well. They're going over a maths problem which from the manner in which the guest is defending the theorem, it must be something he has resolved. What is now being argued is the context of the solution: while is does give an answer, to reach that point is laborious. What Will has managed to do is find another solution which would make it much easier to obtain the answer. From the expression on the guests' face, we can clearly see him being disheartened because someone else has managed to do [better] what previously only he had. As he leaves, we see the little smirk on Lambaus' face. From this little scene we understand the nature of the Mathematical field: it is absolutely competitive. It's not about collaborating with others in order to obtain a Scientific certainty; the only objective is to try and find the solution yourself so that you can be admired by the rest of the community.

Shaun is perfectly willing to spend as much time as needs be to ensure that Will talks first. We can see that Will is also realising this and so out of no where begins to tell, what ends up being, a very funny joke. After that, Will discloses that he's met a woman; how it's different from his previous encounters with them is that this one is so different. Naturally this sounds to Shaun like a good thing and after asking when they'll meet again, he's surprised to hear that Will hasn't called her and thinks he hasn't got a clue. But after hearing Wills' reasons for keeping the status quo, he begins to explain why his "philosophy" is a fallacy. The mistake many of us make is that we spend too long waiting for that perfect person. And when like Will we do manage to find them, we think of excuses as to why it could never work out - there must be something wrong because we don't deserve them! Shaun explains that it's the imperfections which are the actual attraction: we love a person because of certain characteristics that they only have and no one else. If you can realise that you will begin to love that and not the qualities that is shared with everyone else. However, this reasoning slips out from Shaun that those memories have been with him for two years. After his explaining, Will asks about him considering "re-marriage" and Shaun sternly tells him that his wife's dead; to marry will betray her memory, their marriage. Will throws back Shaun's whole argument about not getting to know anyone else because, evidently that's how Shaun is living his life by. He realises that he now can't refute Wills' argument here and luckily enough for him, the session is over.

However, Will does take on board Shaun's advice about pursuing Skyla and proceeds to meet her. It's really awkward because he's left it really late and seems like it's over. Luckily it's not about what happened, but Skyla has revision to do. Being the genius that he is, not only does Will know the problem Skyla is working on, but he works out the whole formula without any materials in a very short period of time. They then spend the rest of the afternoon at the races, having a good time. While there Skyla does the natural thing when a person is interested in another: they want to know about their family. Although Will doesn't show it, the question throws him off and makes a plausible concept of how he manages to have a big family. Luckily for him he manages to handle the burden of proof by reciting (twice!) all twelve names of his 'brothers'. What he can't get away from is Skylas' request for her to meet three of them.

To get better acquainted with Shaun, Will reveals that he's read his book and wants to know more about what Shaun is doing. After seeing how much influence Shauns' wife had on his life, Will asks a very reasonable question about how different would his life have been if they never met. Shaun acknowledges the pertinent question and explains that no matter what decisions a person makes in life, there will always been good and bad consequences to it. However, if the decision was for something a person wanted, then the good will always outweigh the bad; in Shaun's case, not regretting a single moment being with her, despite the suffering he's had to endure from the time of her death through to what currently is perceived, something which he will continue to endure till his death. The next most important question to Will is working out how to know the right one. I think we were all surprised that Shaun managed to give an exact date! However, how a person manages to do something like that is when they have a symbol which they can use to associate with an instance. In this case, the baseball game. As Will is a baseball fan, he enthusiastically gives Shaun more attention as he begins to tell the story about that night. It was while waiting for the game at the bar (the same way Will met Skyla) that he meets Nancy. Shaun continues to talk about the game and both of them are loving the re-enactment, Will asks if Shaun too charged the field and after is shocked to find out that Shaun wasn't even there: he was watching the game with Nancy at the bar. Will knew that if he was in that position, his friends would never had allowed him to miss the game, especially if the reason was to be with someone who he'd just met. The corny line Shaun throws would never have convinced Will; it's when Shaun talks about the look his eyes gave his friend made Will realise that yes, something like that would absolutely convince another of the seriousness of the moment, without the need to say anything. Making that decision was infinitely better then all the apparent negative circumstances that were derived from it: it made everything that would have normally been deemed as significant moments seem inconsequential. How does Will appreciate this new found wisdom? He runs into bed with Skyla! She convinces him the only way a man can be affected to meet his brothers. Will finally gets them to all hang out.

They're all hanging out and true to form, they accommodate Skyla the only way they know how. Will could have begun to regret the idea of bringing her, but they're all surprised when she manages to tell a joke that is more vulgar then anything they'd be able to tell! That firmly allows her to be accepted by all. As they're leaving, Will casually asks to borrow the car to drop Skyla. Chucky naturally thinks he's continuing to joke, but gets a rude awakening when Skyla tells Will that she still wanted to meet his brothers. Chucky just gives a look that implied "dude, what the hell you doin?" but obviously Will doesn't meet his gaze and just avoids the whole situation.

Shaun loved the joke about the aeroplane because he's now telling it to a bartender, who we can work out is his friend. The earlier presumption about Lambau could be substantiated in this scene. My contention was that he is a man who only socialises with others of a similar calibre, and there were many hints in his actions which confirm this. As he arrived he seemed uncomfortable, and even after shaking the bartenders' hand he still seemed out of place. When Shaun asks what he would like to drink, he replies "a Perier"; a person who would attend an establishment like a bar would know that the vernacular is uncommon there. When the bartender jokes to Shaun about paying his tab, it confirms the idea that unlike Lambau, Shaun has no problem hanging out in the 'real world'. What also becomes clear is the reason for Lambaus meeting with Shaun. He's hoping that after the few sessions with Shaun, Will has sorted out all his issues and then can begin 'real work'. Without discussing his intentions with Shaun first, Lambau has already advertised Will to his associates and naturally there's now a huge demand for him. Shaun is naturally shocked by this idea and tells Lambau that now is not the time to discuss Wills' career objectives. This in turn frustrates Lambau because he knows that he too is sitting on a winning lottery ticket (Will) and wants to cash in as soon as (by having him work). However, Lambau can't actually tell this to Shaun and so he thinks up a parable. He asks the bartender of knowing world renowned personalities, and because of their achievements, they joke at the idea of the possibility of the guy not knowing them. But then Lambau throws in the red herring: does the bartender know Gerald Lambau? When he replies in the negative, Lambau becomes confident in knowing how well he has made his argument, and there's a shot of Shaun who also acknowledges what Lambau has done. Lambau proceeds to explain that what he's trying to do is not for personal reasons (although we know better) but that he recognises the potential Will has; that even after all Lambaus' achievements, it has not enabled his status to be on the same level as the legends he mentioned to the bartender. Although credit could be given to Lambau for this concession, we know that he will use Will by colluding his achievements as part of his. This was done earlier in the film where Will had demonstrated his new formula to a Professor, but Lambau was the one who orchestrated the whole meeting, and can claim something like he was the one who identified a weakness in the original work and guided Will to producing something better.

Lambau further demonstrates the superiority of Will by likening him to Einstein. He - like Will - also did not have a formal education, but yet managed to revolutionise Physics. If he chose to act like Will and just had a good time, then the whole world would have missed out. While this would have won most arguments, Shaun immediately responds with his own scenario: he tells Lambau about a guy who had done some major achievements in Maths. Naturally Lambau wants to know who he is, and when Shaun gives the name, Lambau has never heard of him. This would be surprising because as Lambau has dedicated his life to Maths, he would have come across Ted Kazinskys' work because Lambau too would have been establishing himself, and being aware of any competition that would have jeopardised his fame. Shaun asks the bartender, and before giving us the time to assume that he could never know, he responds by saying that Shaun is that actual person! So Shaun has assumed a total new identity, and the reason for someone to do something like that is because they want to be completely removed from having any association with their previous incarnation. Lambau fails (or refuses) to consider this and assumes that Shaun had no guidance, but now that both he and Lambau are there they'll be able to point Will in the right direction. To Shaun this is evident manipulation because if Will really was allowed freedom of choice to do what he wants, then no one should 'tell' what Will should do in life. Now Lambau is blatantly frustrated and no longer withholds his professional demeanour, he's now shouting at Shaun that he wants what is best for Will. There's too much potential there and Lambau is presuming that Shaun is preventing Will from attaining what is rightfully his, but by also including himself as a possible hindrance, he proposes that the both of them keep away from Will and allow him to flourish as much as he is potentially able to. Shaun informs Lambau that he has no personal interest in Will but only that Will should be allowed to decide whatever he wishes to do - with no time frame. Lambau, already frustrated as he is, takes a dig at Shaun by telling him that if that's the advice he gave to himself, then look at how it ruined his career. Shaun too now loses it and no longer bothers to try and justify his views to Lambau. Lambau now completes his overall manipulation of the meeting by disclosing that he arranged an interview for Will.

We now see how Will deals with that manipulation personally by conducting the interview with a prestigious company vis a vis Chucky! Clearly he has not been given any instruction on how to conduct himself at the interview, and being in an environment that oozes money, tries to hustle the interviewers for money! Chucky isn't so naive as to do it blatantly and blow his cover, and it's a nice scene which demonstrates that although he may not be articulate and profess his deep analysis of life, he does understand how the world works and this is a significant aspect which will be demonstrated later in the film.

We see that Will is hanging out with Skyla...and that he's bored. He wants to help [or really] do the work for Skyla so that once it's out the way they can actually start hanging out and having a good time. However, this is the first time that she puts her foot down and [politely] tells Will that it's not a matter of just doing the work, but that she has to understand how it's done so that when she is tested, she is able to replicate the work. Will remains unconvinced because he only sees the immediate implication of that reasoning: it will mean them potentially staying there for the rest of the day. For Skyla that means if that is the case, then so be it. However, she is not completely selfish and only considers her needs; she acknowledges his boredom and knows that if it is not addressed then their remaining time together would be uncomfortable for the both of them. What she chooses to do is try to ascertain where that boredom is being derived from. So far Will has demonstrated his knowledge on a variety of subjects, without the need to resorting to materials. Therefore he must have "a photographic memory"? However, this is a notion that Will has never contemplated before because he doesn't provide an elaborate explanation for his thought-process. We know that Will would have nothing to hide from Skyla because he gave her the coffee/caramel analogy beforehand. Then she wants to know if he's studied what she currently is, and can't conceive of the idea that he has - even if it is only a little! She rightly assumes that the idea of him doing that is crazy: anyone who studies has a reason for doing so: they want to utilise the knowledge acquired in a way that will help them professionally. But Will has shown that he has no interest in pursuing a career, and what is especially important is that Skyla carries on their relationship whilst knowing this.

She knows she cannot assume that Will is just lazy because she is in an environment where there are many others who are extremely clever, but yet they have to still study to get to that point. But she has witnessed Will who is also not only on the same level, but can easily surpass everyone. What is so shocking is how effortlessly he can do this. Instead of reprimanding Will for not making better use of his talents, she is genuinely interested in what makes his mind work - what makes him tick. Will asks if Skyla plays the piano and we assume that he is deflecting, just like how he does with his friends. Skyla pursues and this time Will is also prepared to try and explain. He gives examples of the greatest pianists who got to the stature of where they are now acknowledged in the symphonic world; not because of the lifetime of hard work to get to that stage, but that they were born with that gift: composition came naturally to them. Will acknowledges that he has limitations in many aspects of life, that there are many others who would be considerably better then he. Only in the aspect of Academic knowledge can it be comprehended to him, for the sole reason that it simply can be. After this confession, Skyla plants a passionate kiss on Wills' lips: her form of acknowledgement for accepting and loving Will for who he is. Just from the look in Wills' eyes shows that he can't believe someone is loving him for who he is, and not just a body to be used.

When Skyla wakes Will up and asks him to move with her to California, I think all the guys watching wouldn't rebuke Will when the idea of commitment suddenly arises! At first Will can't believe what she is asking and wants her to actually realise what she is doing. When she knows what she is asking, then projects his own fears on her by wanting to know how she's decided that Will could be 'the one'. Love can never be intellectualised, and Skyla feels that the only reason that needs to be given is to let Will know that she just feels he's the one for her, and that should be a good enough reason for him. Instead he gives his reasons for not being able to go with her, and she knows that they're probably the worst reasons ever. She doesn't entertain the idea of him elaborating those reasons, but simply wants to know if he loves her or not. When he responds in the affirmative, then she can't understand why he's not willing to come because the idea of loving another is pretty much the sole reason for two people wanting to be together. If it is not that, then the only other factor could be that he's afraid of going, and she wants to know exactly what it is. To Will, that idea is preposterous because when has he ever shown any fear to her? While it may be necessarily true that Will is not afraid of being physically assaulted, this fact may be derived by his childhood experiences of being abused. Once a person is confronted by such an experience, they know what to expect not just in terms of physical scars, but all the emotional side affects associated with it as well. However, for the first time in his life, Will is being faced with something completely new: the concept of having an attachment with another who he could end up loving.

Skyla questions his lifestyle and sees it from her perspective, and she thinks because he's so comfortable with the way things are; any significant change to it is sacrilege. Will responds by stating she has no right to question his world because she hasn't lived it, therefore she can never understand what it means to him. Surprisingly he then makes the same error she just has by assuming her perspective on life: that she just wants to marry a "rich prick... and sit around with other trust fund babies and talk about how you went slumming too once". Now Skyla is outraged and focuses on a dichotomy of Wills: on the one hand he has no career aspirations and the wealth that would be derived from that path. Yet any faults he finds in others stem from money being the core reason behind it. He could easily decide to go with her to Stanford; have no problem enrolling in a course on the back of recommendations from Shaun and Lambau; both graduate at the same time and make a life for themselves - together. Skyla tell him that money has never been an issue with her, and in fact tells Will for the first time that it was inherited from her Fathers' death. She couldn't care less about status in any form; she's just trying to deal with life as best she can. Her anger is from the fact that not really about the idea of Will going with her or not, but the fact that he's too afraid to confide in her when they are supposed to be in a relationship. She's not trying to make herself appear better then him for how she is handling the relationship, only that he's not completely honest like she has been: "what about your twelve brothers?"

When Will prepares to leave she's not having that either, and she wants to sort this issue out. Finally Will has no proverbial place to turn and just unleashes everything: "This isn't a scar, it's a stab wound; you don't wanna hear how I had cigarettes put out on me when I was a kid... you don't wanna hear that shit Skyla!"

"I do, I want to hear that so I can help you"

"Help me what the f&^%, do I have a sign on my back that says 'save me' do I need that"

"I want to help you because I love you!"

"Don't bullshit me!"

How can Will comprehend the idea of someone wanting to be with him after knowing everything he's just told her? As a child he's been moved from one foster home to another; not because the new family or anyone else who was in charge of Will were concerned or cared about what happened to him, but simply to avoid authorities, they done what was needed to in order so that everyone can get on with their lives. Skyla is not asking for convenience; that she wants to help because it's her job; only that she is willing to accept all of Will on the condition that he love her in return. Will, horrified of this ultimatum, looks directly at her and tells her he doesn't love her. The manner in which he has just told Skyla leaves her completely devastated.

Will's in the Professors' office, and it can be seen from his expression that he's dejected. Instead of acknowledging this, the Professors' assistant is lecturing Will about being grateful "for the opportunity" and just basically having a go at him. The frustration in Will is starting to come out, but what is funny is for all the dedication the assistant is giving to the Professor; in return Lambau treats him little more then his own personal assistant: he's relegated to simply making coffees or attending to any other of the Professors' needs. When Lambau starts to go over Wills' work, Will starts to drop subtle hints about growing tired with his role. He no longer wants to go to Lambaus office for the meetings, and the exasperated look he gives when the Professor is just concerned of the "embarrassment" at the possibility of the work being wrong... Will finally wants out. Lambau is naturally upset when he realises that Will didn't attend the interview, but that feeling was derived more from his perspective rather then what Will has potentially missed out on. Lambau thinks that he's trying to fulfil Wills' potential, but Will throws it right back at him by telling him that he's not even remotely close to his level! "Do you have any idea how easy this is? It's a f&^%ing joke!" Will has categorically stated that for all of Lambaus intent and purposes, he's actually failing miserably. As a person who's never had the opportunity of exercising patience because of all the experiences he'd gone through in life, Will is frustrated with Lambaus incapacity because had he been as capable as Will, there wouldn't have been a need for Will to waste his time doing Lambaus' work for him. After rescuing the work from being completed burnt, Lambau also loses his emotional composure and openly admits at not being capable to do the work; but he is one of the few people in the world who is at least able to identify that in the first place. It's pretty certain that all his life, Will has been told that he's worthless or at the very least, made to feel that way. But now, in very sincere terms, the Professor is telling him that now knowing of his existence has also brought sorrow to his life - but for completely opposite reasons everyone else had also claimed!

Will goes back to his friends' house and again the same scenario is played out every time he does this: the time spent with them is completely trivial. The only possible explanation for this is that Will enjoys the camaraderie between them. What is completely surprising is the following scene, where Will is evidently at a very important meeting. Surprising because Will had told the Professor to not arrange any more interviews, and with Will's decision for the other interview (not be present personally) this really is a surprising move. But we get to understand what has transpired with Wills' opening question "So why do you think I should work for the National Security Agency?". Will isn't so naive to not know what it could mean to work for them, and having the opportunity to do so, Will is there to see what they have to say for themselves. When Will realises that essentially all he will be doing is "code breaking", he's pretty much given up on the idea of working for them. Instead of just walking out or not giving a reason as to why he doesn't consider a career with them, he respects them enough to deliver one of the many monologues in the movie:

Say I'm working at the NSA and someone puts a code on my desk, something no one else can break. Maybe I take a shot at it and maybe I break it, and I'm real happy about it because I done my job well. But maybe that code was the location of some rebel army in North Africa or the Middle East and after they get the location they bomb the village where the rebels are hiding; 1500 people who I never met, never had no problem with suddenly get killed. Now the Politicians are saying 'Oh send in the Marines to secure the area cos they don't give a shit; It won't be their kid getting shot, just like it wasn't them when their number got called, cos they were up touring the National Guard. It will be some kid from Southy taking shrapnel in the ass. He comes back to find the Plant he used to work at got exported to the country is just come from, and the guy that put the shrapnel in his ass has now got his job cos they work for 15 cents a day and no bathroom breaks. Meanwhile he realises that the only reason he was there in the first place is so that we can install a government that would sell us oil at a good price.

And of course the oil companies use skirmishes to pump up the oil prices

over there, it's a cute little benefit for them but it ain't helping my buddy at 2.50 a gallon. They're taking their sweet time bring back the oil of course; maybe they even took the liberty of hiring an alcoholic skipper

who likes to drink Martinis and play slalom with the icebergs'. It ain't too long before he hits one, spills the oil and kills all the sea life in the North Atlantic. So now my buddy's out of work; he cannot afford to drive so he's walking to the job interviews which sucks because the shrapnel in his ass is giving him chronic hemeroids. Meanwhile he's starving because every time he tries to eat, the only lunch time special is North Atlantic Scrad with Quake steak So what did I think? I'm holding out for something better. I figure f^&T it while I'm at it, why not just shoot my buddy, take his job, give it to his sworn enemy, hike up the gas prices, bomb a village, club a baby seal, join the hash pipe and join the National Guard - I could be elected President

We are subconsciously prepared for Shaun's response because while we witness Will speaking, we see that he is no longer in the NSA building but in Shaun's office. After hearing the deeply philosophical reasons for declining the job, you would expect Shaun to at least take a moment to comprehend everything that's been said, and try and de-construct the views behind each point. However, Shaun immediately throws everyone off by asking Will "Do you feel alone Will?" Where did that come from? However, this simple question encompasses the significant difference between Shaun and Lambau. When Will had his quarrel with Lambau previously in his office,

Wills' frustration was derived from the sense of futility: why bother doing the Maths if no one can appreciate it anyway? The Maths was supposed to be a form of expression that would enable Will to express himself in a certain manner: verbose explanation is not required because everything that needs to be said is encapsulated within simplistic formula. But Will was told that only a few people in the world would have been able to understand what he was to produce, and instead of the feeling of joy or any other associated senses, this would only come from people knowing *who* produced it, instead of *what* they actually produced.

    Will had turned to Maths because he was never allowed to express himself verbally. When he did resort to linguistic power, it was only so that he could put down the other without the need to do so physically. Shaun had never tried to test Will the way Lambau had done, and perhaps this is the biggest reason why Will has been drawn to him instead of Lambau. Will also was completely surprised by Shaun's first question, but what he's attempting to do is make Will himself understand where the sense of frustration is coming from. Shaun explains what he means by "soul mate" and Chucky isn't that, because the notion entails that they have to question and not simply appease. Shaun tries to push for an answer which worries Will, but he comes back with plenty of names that have challenged the Intellectual community. By reading their works, Will would have attained the satisfaction his mind would long for. But Shaun continues to explain that this is actually a form of limitation: Will would be unable to grow because he is only learning and not contributing. Perhaps this is also a reason for why Will had chosen Maths in terms of intellectual pursuits: formulas are being created all the time and there will always be opportunities to collaborate with others in order to continuously improve. However, Will would not even do this because instead of collaboration, he will only do what is needed. This personal attribute has been identified by Shaun as a produce of Wills' life experiences; his Philosophy being: "you'll always be afraid of taking the first step, cos all you see is every negative aspect down the road". Will thinks that Shaun is having a go at him for not taking the job, but unlike Lambau Shaun couldn't care about it: he understands that Will also really knows this. Instead of trying to obtain an indirect advantage from Will, Shaun categorically states what he's trying to do for Will: make him comprehend this particular moment in his life. At his age, parents the world over would be doing everything they possibly could to enable their children to have what Will is being offered. But maybe this explains Wills' particular rebellion: he's being this way because of the very reason of not having parents with him now. Other then himself, there is no motivation for gaining a sense of achievement. Having the best job or the greatest qualifications would mean absolutely nothing to the only three people in Wills' life because all they care about is having a good time. Wills' solace up to this point in life has only been with them, and so if it means stifling any real sense of progression for Will in order to maintain their friendship then so be it. Will tries to defend himself: "I didn't ask for this", but this isn't good enough for Shaun and he hits back "No, you were born with it. So don't cop out with 'I didn't ask for this' ". Will is being torn on the one hand by trying to just live a simple life with his friends, but he can't deny his nature because he has and always will look for ways to learn. The excuse about being a brick layer as "honourable" is lamentable to Shaun, who tests the argument with a question: "you could have been a janitor anywhere in the world. Why be one at the most prestigious technical college in the whole world? Why complete formulas in the middle of the night that only one or two people in the world can do, and then lie about it? Because I don't see a lot of honour in that Will".

The gauntlet has been laid down. Will has managed to find an excuse for all his actions thus far; how will he respond to this? Like a God send (humour is intentional), Will suddenly decides to become a Sheppard!

    Shaun's had enough and despite Wills protest of "being friends" he tells him that nothing continues if Will doesn't become serious. From the moment Shaun was introduced in the film until now, we have seen that he is a man carrying a lot of emotional baggage. Throughout we've been given little pieces of information as to why that is. Now Will confronts him with it by labelling him a "burn out"; now it's his turn to make Shaun take a hard look at himself and how he's dealing with life. The reason why Shaun is a burn out is because he's holding on to the memory of his beloved life, and her loss is also causing Shaun to lose touch of life. Will is perhaps the first person to tell Shaun that instead of giving condolences, he needs to be told that 'it's over! you need to move on with your life'. Of course it's a "big hand you lost" but Shaun just has to deal with it. Shaun couldn't let Will deflect the real issue and that's "what do you want to do?" Will too needs to answer this otherwise he also will not be able to move on in life. When Will can't respond, Shaun's exasperated and tells Will that if he continues to live by the philosophy of profound arguments, then the core issues will never be resolved.

"F£%% you"

"You're the Sheppard!"

It would have been unrealistic if Will had just completely walked out on Skyla. It's finally at the stage where she is getting ready to leave for Medical school, but Will couldn't bring himself to say goodbye in person. To try and make Skyla feel better, Will tells Skyla that he's going to loads of fancy job interviews, in the hope that she would understand his reasons for not going. But she tells him that whatever he does for a living won't matter to her (in other words: she will accept Will for whatever he is - as long as he is with her). Will takes the hint but can't think of a response. In her last hope, Skyla tells Will she loves him and it's just heart breaking that Will lets her go. They're both thinking of each other up until the plane leaves, and then it' business as usual for Will. But the first change is Will not meeting Shaun again. Perhaps he's worried that if he went, talking about Skyla might mess him up. But after their last meeting, it seems more logical that Will feels that Shaun too has given up on him, instead of being like Wills friends and just ignore anything serious. Lambau is not happy about Wills absence because he feels that even though he may no longer be able to get what he wants, he could still go around via Shaun. If Shaun could no longer control Will, then Will has now just become collateral damage. Luckily Shaun can see what Lambau is doing and must have resorted to deep wisdom to make him change his mind.

Will and Chuck are taking a well deserved break after another hard days work. As Chucky is fond of Skyla, he naturally asks about her. Will tells him that she left, but the really surprising thing is that it has already been a week! You can just feel the silent shock from Chucky when he finds out how long it's been. As the two of them pretty much spend every day together, he can't work out why he's only found out now. As he's never intruded on Will's life before, he doesn't pursue the matter. Instead he tries to divert the subject and ask about the other major development in Wills' life: job interviews. Will tells him he just doesn't like the idea of working in that environment where he will only feel used. Chucky responds that the only thing that matters is the money - if you have that you can go anywhere [better] and make a proper life for yourself. Will is surprised to hear this because in his mind, all he wants to do is grow up with his best friend and the both of them can have families and be together. Chucky too has now finally had enough, and perhaps for the first time since they've been friends he's told him straight exactly how he feels about what Will is saying: "look Will you're my best friend so don't take this the wrong way, but if in twenty years time if you're still here, coming over to my house to watch the Patriot games, I'm gonna kill ya; that's not a threat I'll kill ya!" Chucky begins to explain why the thought of Will remaining there is abhorrent to him, and Will cuts him off as to say 'look I've heard this all before: why does everyone think I should be this way just because I have this?' but this isn't just anyone who's now going to tell Will - it's his best friend. His reasons are also different from everyone else: it would make Chucky feel proud to see Will make something of himself.

Earlier I alluded to the notion that perhaps one of the reasons why Will does not want to 'make it' is because he has no one other then himself to do it for. But now Chucky is telling him Will's presumed wrong of him. The reason why Chucky had never said anything is because out of respect for their friendship, he thought giving Will time would allow him to eventually come around. Chucky has witnessed the brilliance of Will coming through, like when they were in the Harvard bar. I also highlighted the contention that while Chucky may appear naive, he does understand how the world operates. The example given was when he was in the interview, and how his actions were a reflection of this. Another very subtle example was at the beginning where they all were hanging out at the bar, and Will "took off" for home when it was only 10pm. Chucky protested but allowed Will to go even though he was given the feeble excuse of Will being tired. Through Wills own admission of what friends should be like when he was talking to Shaun, and how he had a go at Shaun's friends for allowing him to not attend the baseball game, Will too [deep down] would have questioned Chuckys' sincerity by easily allowing him to leave. But perhaps the real reason was Chucky knew the only explanation why Will would do such a thing is to study, and this is a silent form of permission: Chucky was glad Will was doing this. How else was Will gaining knowledge? This too is also an answer to Skylas' query about how Wills' mind works, when they were hanging out together and she was studying while Will was sitting there bored. While everyone else needs to go to an institution and need guidance in order to obtain the knowledge they are seeking, Will is opposite in that he can attain knowledge by himself, but he does need guidance on a personal level i.e. how to conduct himself as a person after having information overload.

We too witness why Chucky is the person he is by explaining himself his views on life. In pragmatic terms, Chucky explains that he knows this is the type of work that he'll be doing for the rest of his life - and he's accepted that. There's nothing Chucky can do at this moment in his life that would provide better opportunities for him. But Will on the other hand has the mental capacity to obtain whatever is required, he just doesn't have the balls "you're too much of a pussy to cash it in" (Authors note: the contrast in genitals is intentional). Will has always heard from others that people are always trying to help others to obtain education; Shaun told Will how his Father worked hard so that he could get an education. But now Will is hearing from someone who he does care about, that basically the reason why everyone does say that is because it's true. Anyone who is 'not doing well' in life would do anything to attain qualifications because, in contemporary times, that is the tool which changes lives. Will asserts that (as they are talking in general principles) Chucky too is making the same mistake everyone else has by assuming he knows what Will wants. Chucky responds by replying it doesn't matter what everyone thinks or knows, it's what he knows and it's telling Will the best part of his day is walking up to his porch and knocking on the door, and Will wouldn't be there. It's a way of saying that because the two of them are so close, it would be hard for them to say goodbye. But Chucky would be OK with it because he knows that Will would have only gone to make a better life for himself. To make sure that Will takes Chucky seriously, he tells him "I may not know a lot of things, but at least I know that". In other words Will may never had or have asked Chucky for advice on anything, but this is the one time in life where Will should listen to his best friend.

After the phone call with Shaun, we'd assume that Lambau would have left the matter with Shaun to deal with: especially as Shaun has been the only person to make a connection with Will. But now we see Lambau in the office with Shaun and he's extremely stressed out. Lambau realises that if Will turns away from Shaun, then Will would also leave him and the possibility of creating exciting Maths formulas would disappear forever. Lambau doesn't hide anything from Shaun any more; he doesn't even care what the two of them might be saying behind Lambaus' back; he just doesn't want Shaun to be the one that will stop Will from working with him.

Lambau's right in his perception that Will is being this way because of personal issues, but wrong in the sense that if Will had no issues, then he'd just be collaborating with him and they'd basically be doing everything that *Lambau* wants. He thinks Shaun is crazy for telling Will that the way he's living his life currently is fine; how can he possibly think this is OK when Will can potentially be living the life of stardom in the Academic scene? Instead of Shaun trying to explain that there is nothing wrong with this type of lifestyle because, ultimately it's Wills' life, Shaun tries to make Lambau understand why he's doing that. We know that Will does want to work on formulas and work out problems no one else can, but he knows that if he does so professionally, then suddenly the whole world will become aware of him. Will would never allow anyone into his life because he won't be able to trust them. The only reason why his four friends have been is because they've proven themselves to him. If Will asked anyone of them to "take a bat to your head they would": they've proved this already in the playground fight. Shaun is trying to make Lambau realise what he's doing right now can jeopardise Will, and Shaun isn't trying to push Will away from Lambau but just trying to protect him. As Lambau is always thinking about himself he still thinks that Shaun doesn't want Lambau to have any success for himself; that if Will turns away from everyone then he'll just be a failure like he has been for all his life. In his desperation Lambau attacks Shaun, because now he knows that he too could have been a great Mathematician just like Lambau. Lambau presumes that the reason Shaun wants to hold him back is because he knows what Will could become, and from jealousy he'll stop him. Now Shaun tells Lambau things he's been hiding inside him for years. Shaun knows that Lambau has always been ashamed of being associated with him 'why does a top Mathematician know some lowly Psychologist?' The only concern Lambau has ever had is trying to impress his "cronies". But despite all the successes he's had in his life and that he's already proven himself to them, Lambau is still scared about doing a work that could be disproved, and thus causing him to lose face in the community. He should realise that everyone does make mistakes, but being happy is only by being content with what you have. Lambau is so arrogant that he thinks Shaun is saying all this because he won the Fields medal! Shaun tells him that doesn't prove anything because he knows what Lambau really is like as a person. Lambau responds by saying that he acknowledges Shaun is a better person then him, but that doesn't matter because what does is the perception the rest of the world has of them. Shaun doesn't pursue the argument about all this being about either of them, his only concern is for Will and he doesn't want him to listen to the poison that Lambau is now subjecting Shaun to just so that he can get what he wants. The only good that has come from this confrontation is that Will has witnessed what's been said between the two, and the surprised expression he has when looking at Shaun is basically saying 'I can't believe you've defended me like that when you don't really know me'. We then see the reason why Will has come to see Shaun: he's realised what he's done by not going to the previous session, and wants to make sure nothing would happen that would make Will go back to prison. Will sees his evaluation and jokes about hoping Shaun hasn't failed him, but the look Shaun gives tells him he doesn't have anything to worry about. As Will is on the defensive, it's now his turn to try and re-establish the connection between the two, and he asks if Shaun too has gone through what Will has. After Shaun tells him about his Father, Will asks about his weapon of choice. Will would go for the one that would inflict the most damage, because it doesn't matter who the person is: they have no right to harm a defenceless person as much as what he had gone through. Shaun realises what Will is trying to tell him, and after gaining Wills' complete attention, tells him that whatever he has gone through in life "what not your fault". In what is the most heart breaking moment in the film, Shaun doesn't just want Will to acknowledge but actually *know* that just because he was born, doesn't make it right for everything that's happened. Through Wills' despair he realises how he's treated Shaun and again for the first time in his life, is regretful for his demeanour.

This causes Will to seriously think about everything that Shaun has talked to him about, and the first thing he decides is what to do with his career. To show his commitment, he walks all the way to the city and presents himself for the interview. We then see some scenes of Boston at sunset, which can be interpreted as showing an end to all of Wills' problems in life, and finally a beginning of a new dawn for him. He's now sitting with Shaun and talking about what he's doing, and he's totally comfortable around him. Shaun wants to only know if this decision is what Will wants, and is happy for him when confirmed. Will is only disappointed when he finds out there will no longer be any sessions, and is surprised when he finds out that Shaun will now be travelling. He is happy when he's told the reasons for this decision, and now their relationship will be cemented forever after

Shaun calls him his "son": through everything they've both attained a symbolic relationship which has been missing from both their lives. The "thank you" they say to each other is embeds significantly more gratitude that can ever be said, but doesn't need to be because it can be derived from the sincerity in their voices as it's being said.

Although Will has re-evaluated his perspectives on life, he shows that he will not abandon everything in order to pursue success: he still will hang out with his friends because they remain the most important aspect. This is reciprocated by them in remembering his birthday, but more importantly, choosing the present for him. The only thing they cared about after finding out where Will would be working is how they can make it easier for him to get there. The fact that they totally disregard all other associations with their decision only enforces their genuineness: what everyone else there would think when they see his car; the reaction from them and what they would say when they find out it was those three who brought it. None of that matters. It also doesn't matter to Will because, although he jokes about it initially, he knows how difficult it would have been for them to purchase a car in the first place, so for him to have one as a birthday present says it all.

After the fight between Lambau and Shaun, it would be presumed that Lambau will walk out of his life now that potentially Will be back in jail. But it does say a lot when Lambau makes the effort to see Shaun and at least try and attempt an apology. Shaun acknowledges the action and just by saying "me too Jerry" lets it be known that it's all good between them. Shaun tacitly brings up the subject of inviting Lambau to a reunion and jokes about buying a drink, and when Lambau laughs it is an indirect joke at the whole idea that he can't believe something so trivial as that was so close to ruining their friendship. No longer would he allow things to get in the way of their time together and invites Shaun to a drink. As they walk off they start conversing without any inhibitions, showing there isn't any malice in their hearts. As Shaun is getting ready to leave, we see Wills' friends coming around to collect him as normal. Something is up when Will decides to leave a note for Shaun instead of meeting him. What is great is we witness how everyone is affected when they realise that Will is leaving them: we go back and witness Chuckys' reaction when he realises that Will really has left, he stays true to his word and shrugs it off and then goes back to tell the others;

Morgan is elated because it finally means he can now sit at the front of the car! And Shaun is only upset because Will stole his line!

What an amazing film.

CPSIA information can be obtained
at www.ICGtesting.com
Printed in the USA
LVHW091511230322
714114LV00005B/116